Monographs on Endocrinology

Volume 17

Edited by

F. Gross, Heidelberg · M.M. Grumbach, San Francisco
A. Labhart, Zürich · M.B. Lipsett, Bethesda
T. Mann, Cambridge · L.T. Samuels (†), Salt Lake City
J. Zander, München

J. Chayen

The Cytochemical Bioassay of Polypeptide Hormones

With 72 Figures and 7 Tables

Springer-Verlag
Berlin Heidelberg New York 1980

Joseph Chayen Ph. D., D. Sc.

Division of Cellular Biology
The Mathilda and Terence Kennedy Institute
of Rheumatology, London, UK

ISBN 3-540-10040-7 Springer-Verlag Berlin Heidelberg New York
ISBN 0-387-10040-7 Springer-Verlag New York Heidelberg Berlin

Library of Congress Cataloging in Publication Data.
Chayen, Joseph. The cytochemical bioassay of polypeptide hormones. (Monographs on endocrinology; v. 17) Bibliography: p. Includes index. 1. Cytochemistry – Technique. 2. Biological assay. 3. Peptide hormones. I. Title. [DNLM: 1. Biological assay. 2. Histocytochemistry. 3. Hormones – Analysis. 4. Peptides – Analysis. WI MO57 v. 17 / WK185 C512c] QH611.C45 574.19′2456 80-13680

Typesetting, printing and binding: Oscar Brandstetter Druckerei KG, Wiesbaden
2125/3020 – 543210

Preface

The cytochemical bioassay system was described in a short abstract in 1971, and more fully, in the cytochemical bioassay of corticotrophin, in 1972. Since then, cytochemical bioassays have been described for several polypeptide hormones, and these assays are already widely used. It is expedient that the subject should be reviewed, as it is in this monograph, by one writer who has had the good fortune to have taken part in the growth of cytochemistry from its early origins to its present position as the basis of possibly the most sensitive bioassay system currently available. However, it should be noted that major contributions have been made by many, both to the development of the subject and to the establishment of the bioassays. The object of this preface is to try to give some perspective to the growth of this subject and to record that the cytochemical bioassay system has been fostered by many outstanding scientists in an atmosphere of remarkable goodwill.

To begin with, there could have been no cytochemical bioassays until cytochemistry had been converted from its rather unsure origins into a precise and quantitative form of cellular biochemistry. This was done with skill and enthusiastic dedication by my colleagues, Dr. Lucille Bitensky, Dr. F. P. Altman, Dr. R. G. Butcher, Dr. L. W. Poulter and Mr. A. A. Silcox, first at the Royal College of Surgeons of England and then, except for the last mentioned, at The Mathilda and Terence Kennedy Institute of Rheumatology, where we were joined by Mr. J. Johnstone. There were many apparently insuperable obstacles, over which others had stumbled, and I can only be grateful to these colleagues for their perseverence, skill and dedication.

When we moved to the Kennedy Institute, D. J. R. Daly (now Professor of Chemical Pathology at Manchester) was responsible for endocrinology at that part of the Charing Cross Hospital (and Medical School) that lay in close proximity to the Kennedy Institute. He appreciated that cytochemistry, as the subject was being developed, could be of particular advantage in studying how hormones influenced the metabolic activities of cells, and even, perhaps, in the bioassay of hormones. Dr. T. B. Binns, then Medical Director of Ciba Laboratories, Horsham, readily agreed to give us a small grant to look into these possibilities, and Mr. N. Loveridge joined us to this end. From such vision, and from a total grant of less than £ 1000 a year, these assays developed.

But my colleagues and I were cellular biologists, not endocrinologists. The transition from cellular biological phenomena to practical bioassays was due largely to the far-sighted understanding and enthusiasm of Professor J. Landon, who saw the potential of our investigations and who guided us in the development of the first bioassay; to the advice of Professor J. R. Hodges; and to the generous collaboration and help of Professor Lesley H. Rees, who put great effort into helping to establish these assays. Indeed, one of the main reasons for the rapid advancement of these bioassays has been the enthusiastic and remarkably generous help we have received from the major experts on each hormone we have studied, and from the leaders of bioassays generally. Thus we had generous help from Professor G. M. Besser in developing the assay of corticotrophin; from him, Professor R. Hall, Dr. K. E. Kirkham and, more recently, Professor J. M. McKenzie, over the assay of thyrotrophin and thyrotrophin-releasing hormone; from Dr. J. A. Parsons and Dr. Joan M. Zanelli with the parathyroid hormone assay, and Dr. J. L. H. O'Riordan in its application; and from Professor Judith Weisz over the luteinizing hormone assay. Dr. A. S. Outschoorn, then of the World Health Organization, gave invaluable encouragement. Throughout we have had encouragement, advice and practical help from Dr. D. R. Bangham, of the National Institute for Biological Standards and Control; I am personally indebted to him for much advice on bioassay generally, as well as for the unstinted supply of samples of reference preparations of hormones which we have received from the National Institute.

I cannot adequately express my gratitude to all these experts, and to very many other endocrinologists and radio-immunoassayists, who helped and encouraged the development of this work. If there be merit in the work, and if the cytochemical bioassay system can help to advance the study of endocrinology, then much of the honour is due to them and to their readiness to welcome this unorthodox approach to their subject. But beyond this, my deep and warm gratitude is expressed to my immediate colleagues, whose names are to be found in our joint publications on cytochemistry and cytochemical bioassays.

My special gratitude is to Dr. Lucille Bitensky who has more than fully shared the trials and tribulations attendant on taking quantitative cytochemistry from its first faltering steps to its present position. Her contributions have been at least as important as my own and she has been largely responsible for developing it as a subject that has practical applicability.

I also wish to record my indebtedness to Miss P. C. Lamb, B. A., for her unflagging kindness and efficiency in dealing with this manuscript, coping with the bibliography, and for trying to remove the more grating errors of language.

My understanding is that these Monographs on Endocrinology are intended as fairly personal records of work done by single groups. Obviously that work needs to be put into some perspective in relation to results and ideas of other laboratories; anyway, no one works in isolation, and any new advance depends on, or is influenced by, the whole body of knowledge relevant to that advance. Consequently the bibliography reflects mainly those publications which I am conscious of having influenced our own work. Finally I must emphasize that the errors, idiosyncratic interpretations and bias must be laid at my door, and at no one else's; they occur in spite of all the generous and excellent advice I have received.

April 1980 J. Chayen

Contents

Chapter 1

Nature of Cytochemistry

A. Definitions and Origins

I. Confusion Between Cytochemistry and Histochemistry

Many equate cytochemistry with histochemistry. Consequently it frequently comes as a surprise to learn that the most sensitive methods at present available for assaying polypeptide hormones involve cytochemistry. Therefore, at the outset, it is essential to appreciate that cytochemistry, or cellular biochemistry, is in fact a very different subject from histochemistry, the latter as generally practised being an extension of histology or histopathology whereas the former is an extension of biochemistry. Although the development of cytochemistry has been reviewed recently (Chayen 1978 a) some explanation of how these subjects differ, and how they relate to the older disciplines, must be given here.

In fact, there are good grounds for confusion between these terms. Under the general title of 'histochemistry' can be included the rather poorly specific staining of chemically fixed, paraffin-embedded sections. As Glick (1967) has pointed out, such staining constitutes the greatest volume of work done and published as histochemistry, and this is identified by most of the scientific community with the whole of experimental histochemistry. Yet Glick has used the same name to describe rigorous chemical analyses of histologically defined samples where the chemical methods are precise and quantitative (Glick 1962, 1963), being based on the highly sensitive microchemical procedures developed at the Carlsberg Laboratories under Linderstrøm-Lang and Holter (e.g. see Holter and Møller 1976).

II. Definition of Histochemistry

Because of this confusion, the present author prefers to accept that there are two completely disparate disciplines. On the one hand there is histochemistry, as practised by most histochemists and as exemplified by the work of Barka and Anderson (1963); on the other there is cytochemistry, as practised by the present author. The essential difference between these two subjects is defined by a fundamental difference in outlook and in aims. According to Barka and Anderson (1963) histochemistry is

> a system of chemical morphology that adds another dimension to histology but which shares the basically static character of the morphological sciences. Its contribution cannot be assessed in the dynamic, physiological terms of biochemistry . . . finite measurement is not the immediate goal of microscopic histochemistry. Deriving its theoretical foundations from chemistry, histochemistry remains essentially a morphological tool.

In contrast, cytochemistry is envisaged by the present author as an extension of conventional biochemistry to the biochemistry of individual cells. It shares with histochemistry a concern to relate biochemical activity to histological structure but the emphasis is on 'biochemical activity', i.e. rigorous measurement of the biochemical activity of the different histological components of a complex tissue or organ.

III. Aims of Cytochemistry

The aims and outlook of cytochemistry are to study the biochemistry of individual cells within a tissue which may contain several different cell types. It is an extension of conventional biochemistry. But it has a further facet, namely that it is a non-destructive method of analysis. Thus whereas Glick's quantitative histochemistry may involve the destruction of the structure of the tissue, as does homogenate biochemistry, cytochemistry tries to retain the structure intact and so to test the biochemical activity of intact cellular systems rather than the biochemical behaviour of these systems after homogenization, and isolation into a foreign medium, have distorted membrane structures and spatial relationships. Thus its outlook is almost the reverse of that of histochemistry, as defined by Barka and Anderson (1963).

IV. Origins of Cytochemistry

Cytochemistry, as defined by the present author, originated in the period 1930–1940 from a number of centres. It owes much to the earlier work of Brachet in Brussels (as reviewed by Brachet 1950), who used a mixture of methyl green and pyronin to demonstrate the localization, and relative abundance, of DNA and RNA in cells. He controlled the use of these dyes by the action of specific nucleases (e.g. Brachet 1954; also Jacobson and Webb 1952). This work demonstrated that there was not a specifically animal nucleic acid (thymonucleic acid) and a specifically plant nucleic acid (phytonucleic acid) as had been thought by biochemists who had found that the animal tissue most rich in nucleic acid, namely the thymus, contained what we now call deoxyribonucleic acid (DNA) whereas the richest plant source of nucleic acid, namely yeast, contained ribonucleic acid (RNA). Brachet (1950) showed that both occurred in nucleated cells, whether of plants or animals.

Much of the spirit of cytochemistry, particularly the need for highly sensitive, very precise, chemical methods which would be sufficiently sensitive to allow biochemical activity to be related to detailed histology, came from Linderstrøm-Lang and Holter in the Carlsberg Laboratory in Copenhagen (see Holter and Møller 1976). This laboratory emphasized the relationship between structure and function and one of the most important concepts of cellular biology, namely endocytosis, came from this group (e.g. Holter 1961, 1965). A practical outcome of their studies was the development of a cryostat microtome (Linderstrøm-Lang and Mogensen 1938) which is now an essential tool in cytochemistry.

Of more direct lineage was the work of Caspersson (as reviewed by Caspersson 1947, 1950). He first developed an ultraviolet microspectrophotometer so that he could measure the ultraviolet absorption spectra of the nucleic acids, and later of proteins, inside cells with the same accuracy as could be achieved when the nucleic acids were extracted from a large sample of tissue and measured in a more conventional spectrophotometer. He defined the optical problems underlying microspectrophotometry and so laid the optical basis of quantitative cytochemistry.

This group at the Karolinska Institute in Stockholm also developed other biophysical tools for measuring biochemically significant factors in single cells, or in small samples. Apart from microspectrophotometry, they showed how soft X-rays can be used to determine the weight of cellular structures (e.g. Engström and Lindström, 1958) and pioneered the X-ray analysis of metal ions in tissue sections (e.g. Engström 1946, 1962). This type of approach was taken up by Sir John Randall's group at King's College, London. They measured DNA and RNA by ultraviolet microspectrophotometry (as reviewed by Walker 1958; e.g. Walker and Yates 1952; Chayen and Norris 1953); they measured mass, by interferometry (Davies et al. 1954; as reviewed by Davies 1958) and particular bonds, such as the phosphate backbone of DNA, by infra-red microspectrophotometry (Fraser and Chayen 1952); the development of scanning and integrating microdensitometry by Deeley (1955; also Deeley et al. 1954, 1957; La Cour et al. 1956) gave cytochemistry its essential tool for quantifying the amount of a cytochemical reaction product in individual cells. Danielli, who was then Professor of Zoology at King's College, London, was concerned with the more chemical and physicochemical aspects of cytochemistry, and initiated the stoicheiometric covalent cytochemical reactions (Danielli 1953).

Mention must also be made of the development of quantitative autoradiography for the analysis of the fate of radioactively labelled compounds and precursors. This was done independently by Leblond in Canada (e.g. Leblond et al. 1948, 1963) and by my late colleague, Dr. Stephen Pelc in London (as reviewed by Pelc 1958). Pelc's work established that DNA synthesis was a discontinuous process during interphase, occurring solely in what he called the S-phase, but not before (G_1) or after (G_2) that period of interphase (e.g. Pelc and Howard 1952; Howard and Pelc 1951, 1953). As yet this technique, and the other optical techniques of quantitative polarized light microscopy and of microfluorimetry, have not been used for the cytochemical bioassay of hormones.

Thus the basis of quantitative cytochemistry was laid by the early 1950s. The only problems to be solved were the adequate preparation of tissue sections and their protection during the reactions for the assay of enzymes and of other reactive groups. The solution of these problems had to wait 10 years; it will be described in Chap. 5. But once they could be solved, the way was open to the assay of the biological activity of hormones, or of any biologically active molecule.

B. Relationship Between Cytochemistry and Biochemistry

I. Material

Fundamentally the practice of biochemistry and the practice of cytochemistry are very similar (Table 1.1). Although it is most common, nowadays, for biochemical studies to be done on an homogenate of the tissue, often followed by differential

Table 1.1. Comparison between cytochemistry and conventional biochemistry

	Conventional biochemistry	Cytochemistry
Material	Thick tissue slice; or homogenate with or without differential centrifugation; or isolated cells	Tissue section 10–20 µm thick; or isolated cells
Reaction	Optimal concentrations of substrate, co-factors and optimal pH	As for biochemistry. But also optimal concentrations of chromogenic and trapping reagents
Specificity	a) for specific substrate b) control by use of specific inhibitors	As in conventional biochemistry
Reaction product	Soluble chromophore	Chromophore precipitated at, or close to, site of generation
Measurement	Spectrophotometry or spectrofluorimetry of chromophore in solution	Microdensitometry (or microfluorimetry) of precipitated chromophore
Size of sample	10^6 cells	1 cell

centrifugation to isolate the main cellular organelles from each other, this is not invariable; particularly in earlier times many biochemical analyses were made on thick tissue slices which might have been about 200 µm thick. The precise thickness was not important, provided it allowed diffusion of reactants into the tissue, because the activities measured were related to unit weight of the slice, or to its protein content. It is to this type of biochemical analysis that cytochemical procedures are most akin in that they involve the use of tissue sections. However, the tissue sections used by the cytochemist must be prepared with much greater precision than was required by the biochemist. The sections of fresh tissue are prepared by chilling the tissue to harden it sufficiently that it may be cut on a microtome at low temperature (−25° C). As will be discussed later, this must be achieved without causing ice to form in the tissue because the formation of intra- or extracellular ice will invalidate many of the studies which the cytochemist may wish to make; it certainly makes it impossible to assay the effect of hormones on the cells of such tissue. Moreover, the thickness of the cytochemist's sections is of critical importance and special steps have to be taken to ensure constant thickness (as discussed in Chap. 5).

II. Reaction

In performing the biochemical reaction the biochemist is concerned to ensure that the reaction is done at the optimal pH and that all the reactants are present at optimal concentrations. The cytochemist is no less concerned to ensure that all these factors are optimal for his reaction; this is of more immediate practical consequence to the cytochemist who is quickly made aware of the fact that the optimal pH for an enzyme derived, for example from the liver, may not be optimal for the same enzyme in another tissue. This is not surprising because it is now well known that the pH characteristics of enzymes on solid surfaces, or matrices, are very susceptible to variations in the nature of those surfaces (e. g. McLaren and Packer 1970; Hornby et al. 1966, 1968; Filipusson and Hornby 1970). Moreover there is evidence that the isolation and purification of an enzyme can appreciably change its chemical characteristics (e. g. Mahler 1953; Siekevitz 1962); even the medium into which it is isolated can alter its specific activity and pH optimum (Altman et al. 1970). This growing awareness that enzyme activities can be significantly altered by the structure in which the enzyme is embedded was well expressed by Bittar (1964):

> In any case it is well recognized that structure and function are inseparable at the molecular level. Since the cell is a multi-phase system and the bulk of the available knowledge is based on enzyme activities in the test tube, the next line of attack is obviously one of charting enzymes and pH relationships in terms of regional distribution in the cell.

(Also see Roodyn 1968.) One of the particular advantages of cytochemistry, as developed by the present author and his colleagues, is that it is a non-disruptive multiphase biochemistry which can be applied to individual cells in a complex tissue (as discussed fully by Chayen and Bitensky 1968).

For all these reasons it behoves the cytochemist to check the pH optimum and the optimal concentrations of reactants for each enzyme in each tissue studied, and not to rely on biochemical data established for the enzyme isolated from a particular organ, or organism, in which it happens to occur plentifully.

III. Precipitation

1. The Need for Precipitation

The real difference between conventional biochemistry and cytochemistry is that for the latter the reaction product must be precipitated, and the precipitate must be as close as possible to the site at which the reaction occurred. The conventional biochemist prefers to have a soluble reaction product which he can measure readily in the reaction medium. He aims at measuring the particular biochemical activity in unit mass of his tissue: for example, to show that the liver of female rats of a named strain contains so many units of this biochemical activity per unit mass (or per mg tissue nitrogen). If he is concerned with the cytological distribution of this biochemical activity he first separates the components of the cells by differential centrifugation. He can then say that most of this activity was found, let us say, in the mitochondrial fraction although a certain percentage occurred in the lysosomal pellet. However, as will be discussed later, each tissue is not composed of a single cell type. The cytochemist, who needs to define the biochemical activities of each cell type within that tissue, must ensure that the reaction product is precipitated as it is formed. Under these conditions he will be able to say, to take one example, that the cells of the zona glomerulosa of the adrenal gland have x units of activity; those of the zona fasciculata y units; whereas those of the zona reticularis have z units and are the only cells to show a uniform response to the adrenocorticotrophic hormone. Had he been a conventional biochemist he would not have noticed the response of the cells of the zona reticularis because, as a fraction of the activity of the whole gland, the change induced in them would have been too small to be significant.

The need to precipitate the reaction product involves a technology which is not met with in the other forms of biochemistry. To produce meaningful results the technology must be applied rigorously. Because the methods may be novel to anyone unacquainted with cytochemistry, they will be discussed in some detail.

2. Methods for Obtaining an Insoluble Reaction Product

There are four ways of obtaining an insoluble reaction product. In the first, the reaction product is bound to an insoluble tissue component; in the second (the simultaneous coupling methods) the reaction produces a soluble product which is coupled, or complexed, or trapped by specific ions or other substances included in the reaction medium; for the third (the post-coupling techniques), although the initial reagent is soluble, the result of the cytochemical reaction renders it insoluble but still colourless so that it has to be exposed to a chromogenic reagent before it can be seen or measured; in the fourth it is the biochemical reaction which converts the relatively colourless, soluble reagent into a highly coloured, insoluble reaction product.

a) Binding to an Insoluble Tissue Component

In these reactions the chromogenic molecule becomes bound, generally through electrostatic or covalent linkages (or by other forces as when a lipophilic chromogen is 'adsorbed' into the lipid layer of membranes), to a structural component of the cells. Possibly the best known of these reactions is the Feulgen nucleal reaction. In his chemical studies on what we now call deoxyribonucleic acid (DNA), Feulgen noted that mild acid hydrolysis, which was insufficient to permit appreciable breakdown (depolymerization) of the whole molecule, exposed bound aldehydes. When Schiff's reagent was reacted with these groups, the reaction stained the DNA of nuclei in tissue sections (Feulgen and Rossenbeck 1924). Since the acid hydrolysis is carefully controlled so as to cause minimal degradation of the DNA backbone, the DNA remains at its site in the cell; the Schiff's reagent attaches to the aldehyde groups in the deoxyribose moieties of the pyrimidine nucleotides and so colours the still insoluble DNA (Fig. 1.1). In this way the soluble chromogenic reagent (Schiff's reagent) becomes insolubilized by the molecule with which it reacts.

When the phosphate moieties of DNA are coloured quantitatively by the dye methyl green, the binding is electrostatic. This dye (Fig. 1.2) is of peculiar advantage in that the distance between its two charged groups is equal to the distance between two gyres of the double helix (Kurnick 1949; Kurnick and Mirsky 1949; Kurnick 1950). Thus the electrostatic interaction between the negatively charged DNA-phosphate groups and the positively charged nitrogen atoms of the dye bind this otherwise soluble dye to the insoluble DNA.

b) Simultaneous Capture Methods

These can be of two types, depending on whether the trapping agent is a metal ion or a chromogenic reagent, but the principle is essentially the same in each case. Both are used in the study of enzymatic activity. Probably the most fully studied is the use of calcium ions for trapping phosphate liberated by the hydrolysis of β-glycerophosphate by alkaline phosphatase (Danielli 1953, 1958):

$$ROPO_3H_2 + H_2O \rightleftharpoons ROH + H_3PO_4.$$

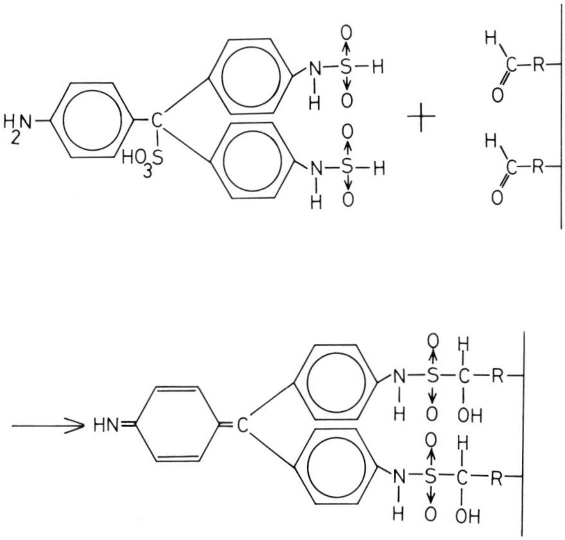

Fig. 1.1. The Feulgen nucleal reaction. Leucobasic fuchsin binds to the disclosed aldehyde groups in the non-solubilized DNA molecule and, in so doing, re-establishes the double-bond system in the fuchsin that restores colour to it

The phosphate so liberated is then precipitated by calcium:

$$xCa^{2+} + yPO_4^{3-} \rightarrow Ca_x(PO_4)_y.$$

In the simplest case x = 3 and y = 2. The solubility product of this calcium phosphate, K_s, is expressed by

$$K_s = [Ca^{2+}]^3 [PO_4^{3-}]^2$$

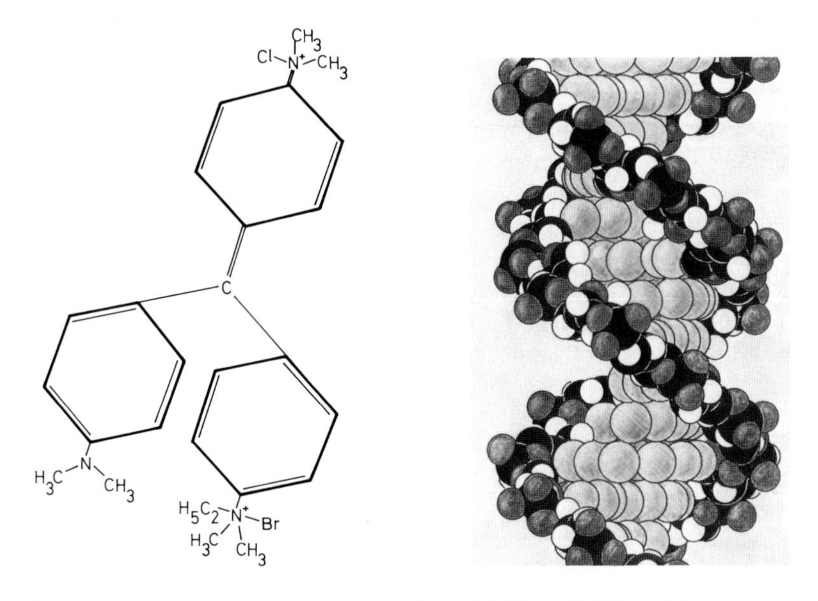

Fig. 1.2. The methyl green reaction with polymerized DNA. The charged nitrogen (N^+) at two sites in the dye molecule binds electrostatically with the negatively charged phosphate groups at set distances along the helical structure of the non-solubilized DNA (shown on the *right*)

where the square brackets denote the concentration of the ion within the bracket. Thus in order to ensure precipitation, the concentration of phosphate ions, which is determined by the rate of the enzymic reaction, and the concentration of calcium ions must be as high as possible. The first is aided by the use of unfixed sections of the tissue, because chemical fixation (as so often used by histochemists) may cause some or much inactivation (see Appendix 1 in Chayen et al. 1973 b, 1975). The second is more under the control of the experimenter. Thus it has been shown that the apparently nuclear localization of some of the alkaline phosphatase activity which has worried many cellular biochemists (e. g. Chèvremont and Firket 1953) is due to having an insufficient concentration of calcium in the reaction medium, so allowing diffusion of the liberated phosphate (Bitensky 1967). Furthermore, Butcher and Chayen (1966) found that the concentration of calcium ions (0.044 *M*) mostly used for this reaction was not only inadequate to trap the phosphate but even allowed 18% of the enzyme to escape into the reaction medium as against 0.7% in the presence of 0.87 *M* calcium. This is a further hazard which the cytochemist has to take into account: namely, that his reaction does not cause the solubilization, and so the loss, of the reagent which he seeks to measure.

The calcium phosphate, precipitated by an adequate concentration of the trapping agent (calcium ions) close to the site at which the phosphate was released, is not coloured. It can be detected and measured (Barter et al. 1955) by micro-interferometry, which measures the dry weight of selected regions in a microscope field (Davies et al. 1954; Davies 1958; Ross 1967). However, at present microscopic interferometers are not ideal for measuring enzyme activity in most tissue sections. This is because interferometry depends on the use of two beams of light, one which passes through the region to be measured, the other traversing the same thickness of the medium in which the section is mounted. In most commercially available instruments, the greatest distance between these beams is about 500 μm, so imposing considerable restriction on the depth within the projected area of a section that can be measured (Chayen 1967). If the shear between the reference and measuring beams can be considerably increased, then it should be possible routinely to measure enzymatic activity by the mass of the precipitated, colourless reaction product.

At present it is necessary to convert the colourless calcium phosphate precipitate into a coloured compound. This is achieved by converting it into cobalt phosphate, which is then changed into the darkly coloured cobalt sulphide. Barter et al. (1955) demonstrated that these changes can be produced without appreciable alteration of the site of the precipitate. Thus a coloured, precipitated reaction product comes to be located close to the enzymatic site.

The alternative procedure is to use naphthol phosphate and precipitate the naphthol moiety by coupling it with a diazonium salt, such as fast blue RR, to yield an insoluble azo-dye. The same requirements of high enzymatic activity and high concentration of the trapping agent apply to such reactions as to those involving trapping by metal ions. The advantage of these azo-dye methods is that they produce a coloured precipitate immediately, and this colour is more amenable to accurate measurement than is the black sulphide of some metals. On the other hand, in these simultaneous coupling azo-dye methods two reactions occur simultaneously, namely the enzymatic hydrolysis and the coupling. This causes no bother when alkaline phosphatase activity is assayed because the alkaline pH which is essential for optimal enzymatic activity also gives an adequate rate of coupling. However, difficulties are met when acid phosphatase activity is to be measured because most diazonium salts will not couple rapidly with naphthol (or its derivatives) at the low pH required for demonstrating fully the enzymatic activity. Consequently, if simultaneous coupling reactions were to be used, they would have to be done at a compromise pH which, of course, diminishes the rate of liberation of naphthol. What with the decreased concentration of the naphthol (in unit time, around the enzymatic site) and with a decreased functional concentration of the trapping agent (because its rate of coupling is less at this compromise pH) the whole reaction becomes somewhat suspect. For these reasons, post-coupling reactions are generally preferred for reactions where the pH for optimal enzymatic activity is not compatible with the process of coupling.

c) Post-coupling Techniques

Instead of using the simple naphthol derivative, such as naphthol phosphate or naphthol glucosaminide, which on hydrolysis produces soluble naphthol, the equivalent derivative of naphthol AS BI (7-bromo-3-hydroxy-2-naphth-*O*-dianisidide) is used. Although the substrate, such as naphthol AS BI-acetyl-β-glucosaminide, is soluble, the naphthol AS BI liberated by the enzymatic activity (in this case, β-glucosaminidase activity) is insoluble. Consequently the sections are reacted with the naphthol AS BI-substrate at whatever pH produces optimal enzymatic activity. The sections are then exposed to the appropriate diazonium coupling agent at the pH which gives the fastest rate of coupling. In this way a coloured, insoluble azo dye is produced close to the site of the enzyme which produced the initial reaction product; each step by which this coloured precipitate was generated has been done under optimal conditions for that step. Of course, the cytochemist will test that there has been no loss of enzyme and no diffusion of reaction product at each point in the procedure.

It is hoped that this discussion of the problems of trapping will show that cytochemistry must be rigorous not only with regard to those features which are well known to biochemists but also to factors which are peculiar to cytochemistry. In many ways it is a more exacting discipline than biochemistry.

d) Tetrazolium Salts

As reviewed in detail by Altman (1976) tetrazolium salts are weakly coloured, freely soluble salts which, on reduction, become intensely coloured, very insoluble formazans (Fig. 1.3). They are used in cytochemistry mainly for measuring the activity of oxidative enzymes. In many respects this procedure is similar to the use of methylene blue in older biochemical studies in which tissue slices (or a tissue 'brei') were reacted anaerobically with aspecific substrate, such as succinate, under the optimal conditions for the selected enzyme, and in the presence of methylene blue. This dye accepted reducing equivalents and, on reduction, became decolourized. (It would be recoloured by oxygen, so the whole assay was done anaerobically.) The amount of decolouration was measured colourimetrically. Thus methylene blue substituted for oxygen in the normal cellular oxidative pathways. But, of course, the decolourized methylene blue was still soluble so it was impossible to define the different oxidative activities of the various cells within the tissue slice. When a tetrazolium salt is used, the formazan is precipitated close to the site at which it is

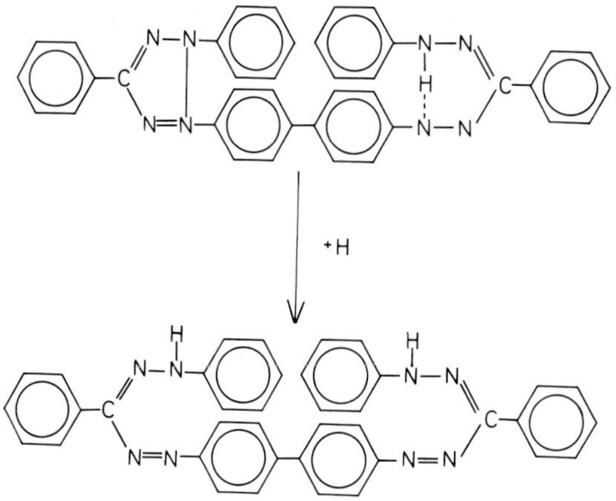

Fig. 1.3. The reduction of the neotetrazolium structure results in the opening of the ring structures on both sides of the molecule (and removal of chloride from neotetrazolium chloride), so converting a pale, water-soluble tetrazolium salt (*upper formula*) to an intensely coloured, very insoluble formazan (*lower formula*)

reduced so that it is possible to distinguish the different oxidative activities of the different cell types within the section. The question of where in the respiratory chains the tetrazolium salts can accept reducing equivalents will be considered later.

IV. Measurement

The aim of cytochemistry is to relate biochemical activity to histology. Consequently the coloured reaction product cannot be allowed to pervade the reaction medium and be measured by spectrophotometric or fluorimetric analysis as is the normal way in biochemistry. Instead, the reaction product must be precipitated and measured in individual cells of each cell type. The fundamental problem was how to measure an inhomogeneously distributed chromophore. Spectrophotometry depends on the chromophore being in true solution at a sufficiently low concentration that aggregates or micelles of the chromophore do not form. The very reverse applies to a cytochemical reaction product. This problem was well recognized by Gomori (1952) who showed that if the reaction section was measured by means of a simple photometer placed above the microscope, considerable error would arise because of the optical inhomogeneity, or non-uniform distribution ('distributional error') of the chromophore. And, of course, the molar extinction coefficient of the solid chromophore is considerably different from that of the chromophore in solution.

This problem was of immediate concern to those who wished to measure the amount of DNA in each individual nucleus. In general such workers reacted the DNA with the Feulgen nucleal reaction and sought to measure the magenta colour. Earlier workers (for review see Leuchtenberger 1958) used a photometer mounted above the microscope and measured only a small sample of each nucleus (the 'plug' technique) and then had to make calculations of the total nuclear content. This was to overcome the fact that the amount of transmission through the nucleus would vary according to the geometry of that nucleus; it did not overcome the distributional error. Other workers produced various techniques for evading this error (as reviewed by Chayen and Denby 1968). The invention by Deeley (1955) of a scanning and integrating microspectrophotometer overcame all these difficulties and opened the way to quantitative cytochemistry generally. This instrument, and the whole question of accurate measurement through the microscope even of a very inhomogeneously distributed chromophore, will be discussed in Chap. 6. But its significance is that the coloured reaction product of a single cell can now be measured precisely. Thus whereas conventional microbiochemistry requires a sample of about 1 mg wet weight (Dingle and Barrett 1969), which represents at least 10^6 cells, the sensitivity of cytochemistry is the single cell (or even part of a cell). This increased sensitivity has contributed greatly to the sensitivity of the cytochemical bioassays of polypeptide hormones.

Chapter 2

Basic Concepts of Hormone Assays

A. Two Ways of Assaying a Biologically Active Molecule

I. General Considerations

There are two completely different ways of measuring how much of a biologically active molecule occurs in a particular sample. The analytical chemist will isolate the substance, purify it and finally weigh it. He can then show that he has so many grams of this chemical, of known purity, per unit volume of the sample. However, certain substances lose their functional capacity when isolated and purified by normal techniques; this was true, for example, of penicillin until special procedures were devised; it remains true for cytochrome P-450, the terminal moiety of the microsomal respiratory chain. For such molecules it is convenient, and sometimes more significant, to measure their biological or biochemical activity. Initially, before column chromatography provided an embarrassment of potential hormones of unknown physiological function (as it were, hormones looking for a job), hormones were measured solely by their biological activity. An Expert Committee of the Biological Standardization Section of the World Health Organization (WHO) would receive results of controlled potency tests done on a standard preparation by various accredited laboratories throughout the world and, on this basis, would assign a unitage to that standard preparation. For example, the committee noted that a particular international reference preparation of partially purified bovine parathyroid hormone was adequate, by a number of criteria including stability, to act as the International Reference Preparation and, based on various evidence, it defined the International Unit for Parathyroid Hormone ('Bovine, for bioassay') as the activity contained in 0.02765 mg of the International Reference Preparation (WHO Report 1975). It is immaterial if this weight of the reference preparation contains non-hormonal matter incorporated to stabilize the hormone. All assays must be expressed relative to the activity present in this weight of the reference preparation. The International Standard of various hormones, such as that for the adrenocorticotrophic hormone (ACTH) or that for crystalline insulin, is kept by one of the WHO's reference laboratories rather as the standard metre is kept (elsewhere) as a yardstick for other measuring devices.

II. Assay of Enzymes

This question of whether we should measure biological activity or the number of molecules, can be put into wider context by considering the assay of enzymes. In general, biochemists report their results as so much activity per unit mass of tissue, or per millilitre of the biological fluid. Let us consider the case of the enzyme, glucose-6-phosphate dehydrogenase. This enzyme is fully active only when it is present as a hexamer, i. e. a molecule containing six apparently identical subunits of the apoenzyme, linked together with molecules of the coenzyme NADP. It readily dissociates into two partially active assemblies each of which contains three subunits. Moreover, each subunit can be in either an active or an inactive form (as reviewed by Chayen et al. 1974 a). To define the concentration of this enzyme as the number of hexameric molecules containing the requisite number of coenzyme molecules at the correct sites in the assembly, would be beyond the capabilities of most enzymologists. Even if this were done, it is possible that conformational changes in the molecules may also be required for full enzymatic activity. Consequently it is not only expedient, but also reasonable, to express the concentration of this enzyme by the amount of enzymatic activity present per unit mass of tissue, i. e. by its biological or biochemical function.

On the other hand, the clinical biochemist may need to evaluate the level of a transaminase in the blood as a measure of cellular damage somewhere in the body. He really needs to know the number of molecules of this enzyme which are present because it is the number, not the activity of each molecule, which increases in the blood as more and more cells become damaged. He may, in fact, measure the increased transaminase activity in the blood simply because this should reflect the increased number of enzyme molecules present (assuming the activity per molecule remains constant). But ideally he would benefit from an analytical method, such as is used by the analytical chemist. A simpler 'analytical' method is now provided by radio-immunoassay which, in effect, measures the number of antigenic determinants, specific for that molecule, which are present in the sample. Yet, while the analytical methods may be ideal for this type of problem, they would be inadequate for measuring glucose-6-phosphate dehydrogenase activity because whereas they can define the number of subunits, or monomeric forms, present in the sample, they will not disclose whether these molecules are capable of dehydrogenating glucose-6-phosphate.

Thus there are two distinct approaches to the assay of biologically active molecules. The first, namely the analytical approach (or 'structurally specific' approach: Ekins 1976), determines in essence the number of molecules present. This can be all important. On the other hand, it can be misleading, particularly in those instances where the molecule may alter its configuration to change from an inactive to an active form, or where slight chemical alteration to the molecule, such as phosphorylation, will have the same effect without the change being recordable by the analytical method employed. The second, namely the functional approach, measures the biological activity of the molecules. It can be argued that this is a more realistic measure of the presence of biologically active molecules but it, too, can be misleading. For example, there are some enzymes which normally exist in the inactive form until the cell or cell membrane is stimulated; the functional assay of such an enzyme would indicate that none was present (in unstimulated cells). Thus it is clear that the two approaches are not in conflict; in some studies one or the other is the better way of expressing the assay; ideally, both are complementary one to the other.

B. Radio-immunoassay

I. General

As Ekins (1974) recorded, saturation assay techniques such as protein binding assay, radioreceptor assay and radio-immunoassay, have 'made an explosive impact upon endocrinology' in the past decade. These methods represent a common analytical approach which has been applied to more than two hundred substances many of which had not previously been capable of assay. Although such assays were first developed for the relatively limited objective of measuring the concentration of hormones in physiological and pathophysiological studies, the value of radio-immunoassay in clinical practice has greatly extended the use of these assays into diagnostic laboratories, particularly those concerned with clinical pharmacology and oncology (Challand et al. 1974).

II. Sensitivity of Radio-immunoassay

Conventional bioassays are usually too insensitive to measure the circulating levels of hormones in plasma which has not been concentrated. Thus the Lipscomb-Nelson bioassay of ACTH (Lipscomb and Nelson 1962) cannot usually measure less than 200 or perhaps 100 pg/ml; the lowest level detected by the intravenous chick assay for parathyroid hormone (Parsons et al. 1973) was 10^{-8} g/ml of blood (Parsons et al. 1975). Yet the normal circulating levels of ACTH may range from about 60 to 5 pg/ml, and those of the biologically active parathyroid hormone have been calculated by Parsons et al. (1975) to be about 10 pg/ml (i.e. 10×10^{-12} g/ml). Consequently the greatly increased sensitivity afforded by radio-immunoassay seemed to be the final answer to the problem of routine assay of circulating levels of hormones. It was at least ten or twenty times more sensitive than conventional bioassays, the radio-immunoassay of ACTH being capable of measuring 10 pg/ml, and it enabled the laboratory worker to handle many samples rapidly, in large batches, as is required in routine laboratories.

But it is becoming increasingly clear that the great increase in sensitivity afforded by this form of assay is still not sufficient. For example, normal circulating levels of ACTH can be as low as 5–10 pg/ml, which is just at the limit, or just below the limit, of radio-immunoassay. Subnormal levels cannot be measured, nor can those levels which apparently may be found by minute-to-minute sampling (Gallagher et al. 1973), and which can be measured readily by the cytochemical bioassay. Even the highly sensitive radio-immunoassay of thyrotrophin cannot measure below 0.5 μU/ml whereas the evidence of Tunbridge et al. (1977) indicates that the levels in a considerable proportion of a normal population are below this value. It is obvious that if radio-immunoassay cannot measure the lower normal levels of TSH, it cannot contribute significantly to discussions on the possible clinical effects of subnormal levels. And similar examples could be adduced for other polypeptide hormones (Table 2.1).

Consequently, although radio-immunoassay has greatly increased the sensitivity with which circulating levels of hormones can be assayed, present evidence indicates that it may be insufficiently sensitive to define the lower levels of normality, and so to distinguish subnormal levels. This stricture applies to many, but not all, polypeptide hormones.

Table 2.1. Relative sensitivity of radio-immunoassay (RIA) and cytochemical bioassay (CBA)

Hormone	Circulating level[a]	Sensitivity of		Units	CBA	
		RIA	CBA		Index of precision	Fiducial limits (%)
ACTH	8–50	10	5×10^{-3}	pg/ml	0.076 ± 0.002	86–115
TSH	0.1–2	0.5	10^{-4}	μU/ml	0.13	75–133
PTH	10–20	75	5×10^{-3}	pg/ml	0.12 ± 0.07	73–136
LH	10	1	10^{-3}	mU/ml	0.12 ± 0.01	–
Gastrin	5–30	1.5	0.005	pg/ml	0.10 ± 0.05	78–128

[a] Based on bioassay, or values calculated from infusion data.

C. The Drawback of the Analytical Approach

I. Problem of Fragments

As has been emphasized, the analytical approach to the assay of hormones aims fundamentally at measuring the *number* of the specific molecules present in the sample. Radio-immunoassay effects this by employing the immunological response between the specific molecule and an antibody raised (and purified) specifically to that molecule. But, in fact, that statement is imprecise. The antibody does not respond to the whole polypeptide molecule; it responds to certain antigenic determinants within that molecule and these can be constituted of as few as three consecutive amino acids.

For simplicity, let us consider the radio-immunoassay of ACTH. This molecule consists of 39 amino acids (Fig. 2.1). The first 24, taken from the N-terminus, are required for full biological activity and appear to be constant in the ACTH molecules found in different species (Ney et al. 1964; Schulster 1974). The amino acids 25–39 are said to give the species specificity (Ney et al. 1964). Thus when human ACTH is injected into an animal to produce an antibody, there will be a tendency to produce antibodies to the species-specific antigenic determinants in this part of the molecule since these are indeed 'foreign' to that animal. Now suppose that the α^{1-39} hormone, circulating in the blood of a human subject, is cleaved (either in the circulation, or in the target tissue, or elsewhere in the body). When a sample of this blood is tested by radio-immunoassay, the antibody will respond to

1 5 10 15 19
Ser-Tyr-Ser-Met-Glu-His-Phe-Arg-Trp-Gly-Lys-Pro-Val-Gly-Lys-Lys-Arg-Arg-Pro-

20 25 30 35 37
-Val-Lys-Val-Tyr-Pro-Asn-Gly-Ala-Glu-Asp-Glu-Ser-Ala-Glu-Ala-Phe-Pro-Leu-

38 39
-Glu-Phe

Variants: -Leu- (a, at Ser 31); -Gln- (b, at Glu 33)

[a] porcine
[b] ovine and bovine

Fig. 2.1. Human ACTH and variant amino acids in other species (Schwyzer 1977)

the antigenic determinants irrespective of whether they are part of an intact α^{1-39} ACTH molecule or in a fragment of the cleaved molecule. Consequently the radio-immunoassay will record the presence of more 'molecules' of ACTH than are actually present.

This danger was recognized early by radio-immunoassayists who therefore tended to use antibodies raised to the N-terminal region of the molecule whenever this was feasible. But, in fact, the same argument pertains. Suppose we have an antibody which responds to antigenic determinants (e.g. 3–5 amino acids) in the sequence 1–24. It (or they) will recognize cleavage products of this region even though such fragments are too small to exert the biological activity of ACTH.

This dissociation between immunoactivity and bioactivity, due to cleavage of the ACTH molecule in the body of the subject, has been reported by several workers (e.g. Besser et al. 1971; Matsuyama et al. 1972). It will be discussed in more detail in Chap. 9. It becomes of even greater moment when the half-lives in the circulation of the biologically active and biologically 'inert' (but immunoreactive) fragments are appreciably different. This seems to be the case in respect of parathyroid hormone which consists of a single polypeptide chain of 84 amino acids. The minimum length required for biological activity is between 1–20 and 1–29 measured from the amino-terminal end. Full biological activity seems to be given by the sequence made up of the first 34 amino acids. Elimination of the first amino acid results in the complete loss of biological potency (Parsons and Potts 1972). If we suppose that the half-life of the immunoreactive fragments, namely the larger part of the molecule, is considerably larger than that of the biologically active whole molecule (or fragment) then it would be possible to imagine a situation where relatively low concentrations of the 1–84 molecule are quickly turned-over in the circulation, giving rise to accumulations of the longer-lived biologically inactive fragments. Thus, until recently, the sensitivity of the radio-immunoassay of this hormone was about 500 pg/ml (500×10^{-12} g/ml) and it was sufficiently sensitive to measure the concentration of 'parathyroid hormone' in the human circulation even though the concentration of the biologically active moiety is postulated to be only 10 pg/ml (10×10^{-12} g/ml). Consequently there seems little doubt that this radio-immunoassay was measuring the accumulation of longer-lived, immunoreactive moieties which, because of their long half-life in the circulation, may bear only little relationship to the concentration of the biologically active hormone.

II. Problem of Different Forms of the Hormone

Apart from the complexities introduced by biologically inert fragments of hormones, it is now becoming increasingly apparent that polypeptide hormones may exist in different molecular forms. One such form is the 'big' hormone: the gastrin molecule may occur as the normal G_{17}, or in big gastrin or even in 'big-big-gastrin' and a 'big ACTH' has also been described (e.g. Yalow 1974). There are indications that a big form of TSH has also been isolated from human pituitary glands (Erhardt and Scriba 1977). Whether these are polymeric forms or aggregates of the simple, biologically active molecule, or whether they are larger polypeptides containing the biologically active molecule, or whether some may be prohormones or act as such by slowly releasing the active hormone into the circulation, is as yet unclear. What does seem reasonably certain is that these big hormones have the antigenic properties of the natural hormone without the full biological activity, in some cases their biological activity is only a very small proportion of that of the natural hormone molecule and indeed some may apparently even lack biological activity.

III. Special Advantage of the Analytical Method

Whether radio-immunoassay measures biologically inert fragments or big hormone, it may record concentrations of the hormone molecule which are higher than those of the biologically active hormone. When the discrepancy between the concentration of biologically inactive and active molecules becomes considerable, radio-immunoassay will report levels of the hormone which are obviously incompatible with the clinical findings. For example, it may record ten times the normal level of gastrin in a patient who has no signs of excessive gastric acid secretion, or elevated levels of thyrotrophin in obviously hypothyroid patients (Krieger 1974; Illig et al. 1975; Belchetz and Elkeles 1976). Such cases are good examples of situations where the analytical approach can be seriously misleading. These are particularly pertinent examples because the assay is being related to the physiological function of the hormone in that patient. And, after all, this is the basic concept of a hormone (as suggested by Huxley 1935), namely that the presence of a hormone is represented by an activity. In a sense, when faced with an over-active tissue such as the thyroid or stomach, we are not primarily concerned with how much thyrotrophin or gastrin is being brought by the circulation to that tissue; in the first place we are concerned to know how much hormone-like activity is being presented to the tissue. Conceptually at least, it would be correct to measure how much thyrotrophin-like activity, or how much gastrin-like activity (assuming that gastrin is indeed related to gastric acid secretion) is present in the circulation. Only secondarily do we need to know if the thyrotrophin-like activity is indeed caused by thyrotrophin (implying an over-active pituitary gland) or whether it is caused fully, or in part, by an immunoglobulin of the long-acting thyroid stimulator type (indicating the possibility of an immunological disturbance). In this type of investigation the analytical approach is valid only if there is a one-to-one relationship between the *number* of molecules recognized by the antibody and the biological activity of each molecule. Clearly, if it were practicable, the functional approach would be the more certain for such studies.

This rather academic argument, that primarily we are concerned with hormone-like activity rather than with the activity of a specific hormone, can have a practical application. Suppose, for example, that pigs or chickens were being given large doses of an oestrogenic drug to improve their meat. In view of the possible physiological (and oncogenic) effects of oestrogen-like substances on the consumers of such meat, it might become of concern to determine the oestrogen-like content of the meat. There is every likelihood that the original drug (which theoretically could be assayed by radio-immunoassay) will have been metabolized by the animal, possibly to more potent oestrogenic molecules. In such a situation radio-immunoassay might well be felt to be wanting; ideally a functional assay might be required which would measure the oestrogen-potency present in the meat, irrespective of the chemical formulae of the molecules which contributed to that potency.

But the analytical approach, which determines the number of molecules which have the molecular (or rather the antigenic) characteristics of the hormone, has certain peculiar advantages. It is now clear that tumours of many types secrete hormone molecules (e.g. Rees and Ratcliffe 1974; Bloomfield et al. 1977). Very frequently the molecules secreted are not the normal, biologically very active molecules, but may contain a high proportion of the 'big-hormone' with little or no biological activity. Consequently a functional assay would not detect their presence. But it is essential to do so because such molecules are becoming increasingly important as cancer markers. They announce clearly that a tumour is present and, by sampling blood from various sites in the body, they can even be used to demonstrate the site of an otherwise undetectable tumour (Rees 1977). The successful, or otherwise, surgical removal of the tumour can be monitored by the total removal of the aberrant hormone molecules from the blood.

At present too little is known of the function of the big hormones which are produced by normal tissues for us to be dogmatic as regards the need to assay these as well as the functional hormone molecules. For example, the gastrin secreted by the duodenum is largely big gastrin whereas that produced by the stomach antrum is mainly normal G_{17}-gastrin (Chap. 12). Consequently a major proportion of the gastrin in the blood of fasting subjects is big gastrin (G_{34}-gastrin) whereas the proportion changes in favour of G_{17}-gastrin shortly after a meal (see Chap. 12). Such phenomena must have physiological significance. Thus at present, in spite of the logic of the argument that hormones represent functional concepts and therefore require functional assays, it is highly important that both types of assay, radio-immunoassay and bioassay, are used. Both types of information are vitally needed. Indeed, it may be that we will finally need to know the specific activity of polypeptide hormones (i.e., the functional activity per unit molecule of the hormone) rather than only their concentrations in the blood.

D. The Need for 'Microbioassays'

The establishment of radio-immunoassays for many polypeptide hormones in the 1960s greatly advanced the assay of hormones. Indeed, about 1968, the present author was informed that there was little interest in a cytochemical study on the action of the adrenocorticotrophic hormone on the adrenal gland because the best outcome of such an investigation would be a bioassay for this hormone and radio-immunoassay had already rendered such assays outmoded. Yet already in 1967, a special meeting convened by the WHO had recommended 'that emphasis be placed on the development of biological microassays which should preferably have a sensitivity comparable with radio-immunoassays, with which they should be run in parallel.' The reason for this recommendation was, as discussed above, that immunoassays 'measure a composite of antigenic activity which is not necessarily related to the bioactivity of the hormone' (WHO Report 1975).

This is an obviously sensible requirement. For example, in the large-scale manufacture of insulin the hormone may be checked by radio-immunoassay at each step in the commercial process. But seeing that such analytical methods tell you well-nigh nothing about the functional potency of the molecules measured, would anyone be willing to inject this 'insulin' into a human diabetic? Obviously the final batch has to be checked by bioassay. But this final batch may consist of a number of smaller batches, some of which may have retained their biological activity while others may not. It might have been helpful to have each checked by a bioassay, but the amount required for conventional bioassays, and the time and expense involved, may render this impracticable.

The need for 'microbioassays' has become more acute with the increasing importance of radio-immunoassay. To measure human pituitary hormones in the blood by radio-immunoassay it is best to use human pituitary hormones as standards in the assay. The WHO, through its Expert Committee, is responsible for commissioning the preparation of such standards and assigning an international unit to them. In the first instance at least, this involves comparing their immunopotency with their biological potency. The amount that can be obtained of such pure preparations of each human pituitary hormone is so small that a considerable proportion would be used if the biological potency had to be determined by conventional bioassay procedures done in a number of laboratories, as is customary in assigning a unitage to a hormone preparation. Clearly there is a need here for a biological microassay system.

These then are the formal reasons for producing yet another assay system. The cytochemical bioassays seem to be capable of fulfilling these needs. But, as has been discussed above, there is also a growing need to relate results of analytical assays to those of functional assays. There is also a realization that radio-immunoassay may not be capable of measuring low-normal and subnormal levels of hormones circulating in man, and it is difficult to apply radio-immunoassay to small animals both because of the need to produce a suitable antibody to the hormone molecule as it occurs in each of these species and because of the small amounts of blood available from such animals. Whereas it is possibly feasible to remove 15 ml from a human subject every 15 min for 1 or 2 h to study the effect of a particular phenomenon, this is not a practical proposition in smaller animals. The advantages of the cytochemical bioassays in such situations, and their applicability to many problems in endocrinology, will be discussed later.

Chapter 3

Introduction to Cytochemical Bioassay

In days gone by it would not have been necessary to detail the essentials of bioassay in a monograph in endocrinology. Today, with the universal emphasis on radioimmunoassay, the present author hopes he will be forgiven for reminding his reader of these essentials.

A. Criteria for Bioassays

I. Basic Requirements

1. Nature of Bioassays

In general, bioassays depend on comparing the effects of the unknown, or test, preparation with those of a standard preparation (as discussed by Gaddum, 1953, who derived mathematical treatments for bioassays). The standard preparation may be an international reference standard obtained from one of the WHO's laboratories or it may be one prepared by a national centre or by the laboratory worker himself, and calibrated by comparison with an international reference preparation. In such bioassays a particular physiological response is measured. For example, it may be the rise in concentration of calcium in the blood of 10-day-old chicks when injected with a range of concentrations of parathyroid hormone (Parsons et al. 1973); three groups, each of five chicks, are injected intravenously with one of three concentrations of a standard preparation of parathyroid hormone in a special vehicle (e.g. 2, 6 and 18 MRC units per chick, 0.6 mg of the partially purified hormone corresponding to 200 MRC units) while three similar groups are injected with different amounts of the unknown, in the same vehicle. This gives a 3 + 3 assay, using a total of 30 chicks. They are bled 60 min after the injection and the plasma calcium is measured. The mean index of precision (λ; as discussed later) of this assay was 0.14. A second example is the Lipscomb and Nelson (1962) bioassay of ACTH. For this the rats are first hypophysectomized. Two hours later the rats are anaesthetized and injected retrogradely up the left adrenal vein, blood being collected from this vein 90 s later for fluorometric measurement of the amount of corticosterone produced. Seven rats were used for each concentration of ACTH tested. In the Sayers assay of ACTH (Sayers et al. 1948), the depletion of ascorbate from the adrenal glands of hypophysectomized rats was measured 1 h after the standard preparation, or the test material, was injected into the tail vein. For statistically significant results, 12 rats were required for each group. Thus a four-point standard response graph and two dilutions of one plasma required 72 rats.

Bioassays of pituitary polypeptide hormones usually require the removal of the pituitary gland to eliminate endogenous production of these hormones. They depend on measuring a specific physiological response induced by the hormone. For example, they measure ascorbate-depletion in the adrenal gland after injection of ACTH, or ascorbate-depletion in the ovary after luteinizing hormone has been injected. They depend on the comparison of like-with-like: the standard preparation must be as similar as possible to the active principle in the test fluid. Because they depend on the response in whole animals, and because the response of different animals can be quantitatively dissimilar ('between-animal' variation) several animals have to be injected with each concentration of the standard preparation, and with each of the test material.

2. Parallelism

The results produced by the different concentrations of the standard preparation are plotted as a graph and compared with those obtained from different concentrations of the test material. The first question to be decided is whether the dose-response graphs are parallel. This often can be seen by inspection; mathematical methods can be used (e. g. Cornfield 1970; European Pharmacopoeia 1971). If they are not parallel it is likely that the test material is not identical with the standard. This may not always be true, but it is best to assume it to be so unless proved otherwise. The fact that they are parallel does not prove that the test and standard are identical, but parallelism is an absolute prerequisite for a bioassay. Given parallelism, it is possible to compare the relative potency of the test material with that of the standard. When three concentrations of test and three of the standard material are compared, the assay is a 3 + 3 assay; 2 + 2 assays may be sufficient when the concentration of the hormone in the test material is sufficiently known that roughly equivalent concentrations of the standard preparation can be made. Most cytochemical bioassays use the 4 + 2 or 3 + 2 (standard + 2 dilutions of the test) form of assay. The essential feature, however, is that the equivalent relative potencies of the test and the standard material are compared as dose-responses (or more precisely as log dose-responses) and that the response produced by different, graded concentrations of each should be parallel.

3. Precision of Bioassays

The concept of 'precision' in bioassay (Borth 1952) is well explained by a simple analogy (for which I am indebted to Professor J. R. Hodges). If a marksman shoots four shots at a target, all falling close together, he may be regarded as highly precise, in terms of bioassay, even if he is as inaccurate as shown in Fig. 3.1.a. By contrast, his efforts in Fig. 3.1.b are highly accurate, in that the mean of all his four shots would lie in the centre ring of the target; but, in fact, none does. He is both accurate and precise if his shots fall as in Fig. 3.1.c.

Thus, as regards bioassay, precision involves the reproducibility of the measurements. But it also involves how much weight should be given to a particular value. For example, when assaying ACTH, is a value of 5 pg/ml different from 7.5 pg/ml or 10 pg/ml? If we consider an hypothetical dose-response graph (Fig. 3.2) which would be completely unacceptable as a basis for a bioassay, this question becomes obvious. In this case the response to the standard, at a concentration of 10 pg/ml, has so much overlap with the response to 1 pg/ml that it would be impossible to say

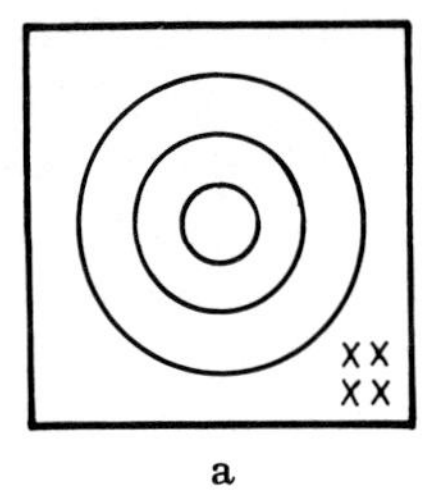

a

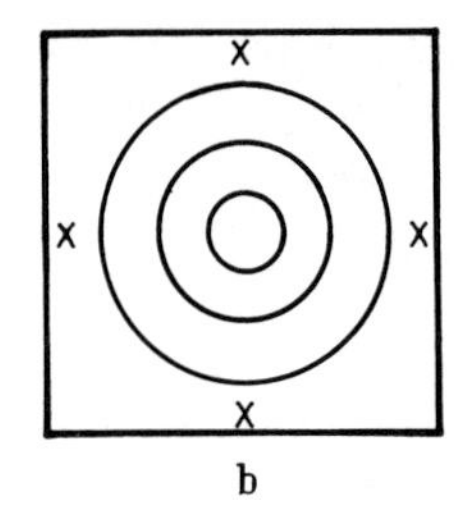

b

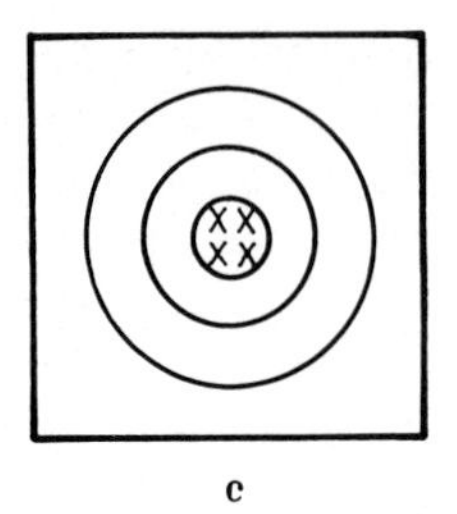

c

Fig. 3.1 a–c. Simple concepts of accuracy and precision. **a** The four shots show good precision in being very reproducible; **b** They are accurate, in that the mean would lie in the centre of the target; **c** Only here are the shots both accurate and precise

that the test fluid contained 5 pg/ml. All that could be said is that this fluid contains between 1 and 100 pg of the hormone per ml.

Thus precision, in terms of the amount of activity found in the test fluid, involves two factors: the slope of the dose-response graph and the deviation of points about this regression line. In this type of graph the regression line can be calculated from the equation $y = bx + c$ where b and c are constants of the equation, b being the slope of the graph. We can measure the standard deviation (s) about the line by subtracting each value of y from its calculated value, y_c (as given by the calculated regression line) in the normal way, namely

$$s = \sqrt{\frac{\sum (y - y_c)^2}{n - 1}}$$

where n is the number of readings.

Since we can determine b from the equation, or by inspection of the graph (the amount that y changes for each unit on the x axis, i. e. 50 units in Fig. 3.2), we have both components that affect the precision with which two readings can be discriminated. This is given by the index of precision (λ):

$$\lambda = s/b$$

In some calculations the value s_{yx} is used; it is identical with s in these equations.

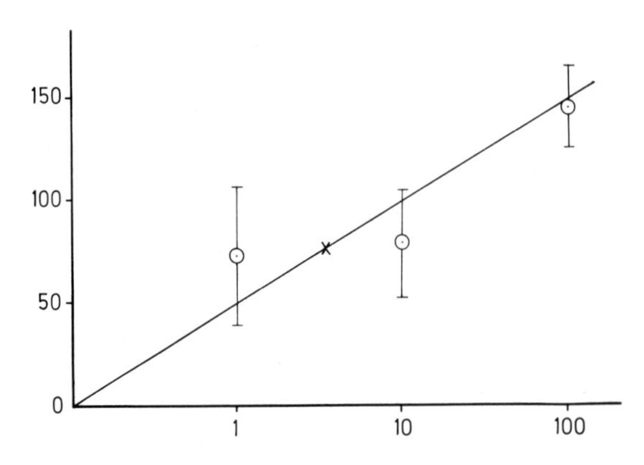

Fig. 3.2. An hypothetical set of results relating activity (*vertical axis*) to concentration of the hormone (*horizontal axis*). The bars show the variation obtained for each set of readings (standard deviation or standard error of the mean). Clearly there is no difference in response to 1, 5 or 10 units of concentration of the hormone

The derivation and significance of λ has been discussed by Gaddum (1953) and by Sayers et al. (1948); both cite previous work concerning this index and discussion of its significance.

An index of precision of better than 0.3 is usually taken to be adequate for a bioassay (Loraine and Bell 1971). Gaddum (1953) suggested that the reciprocal of λ is a better measure of precision ($L = 1/\lambda$) since b/s has the same distribution as t, the probability factor. For an inaccurate assay L would be 2; it would rise to 30 or more for very accurate assays. In terms of λ, these values indicate that a value of 0.5 indicates an imprecise assay whereas a λ of 0.03 would suggest a highly precise assay. Cornfield (1970) pointed out that a λ of 0.3 means, in practice, that the log dose estimated from a single observation has a standard error of 0.3. Thus within the limits of a single standard deviation, that is to 68% confidence limits, a single observation made in an assay with a λ of 0.3 could be in error to $\pm$ antilog 0.3, that is 2 times or a half the value (50%–200%) produced by the assay. In terms of a chemical estimation this sounds shockingly imprecise but, in practice, it is often all that is required of a bioassay. For an assay with an index of precision of 0.1, a single observation has a 68% chance (one standard deviation) of being from 79% to 126% of the correct value. To obtain the 95% confidence limits, twice the standard deviation is required so that we use the antilog of $2 \times \lambda$. Under these conditions the value predicted by a single measurement of the plasma sample would be within 63%–158% of the true value. This value is obtained by taking the antilog of (log 100 $\pm$ 0.2).

The fiducial limits of an assay are usually calculated from at least two observations on the test material and therefore give a fairer indication of how much reliance can be placed on the values recorded. The equations for calculating the fiducial limits of an assay (Gaddum 1953; European Pharmacopoeia 1971) take into account the parallelism of the response to the test and the standard. They compare $2 + 2$ or $3 + 2$ or $3 + 3$ assays and give the 95% confidence limits of the particular assay which is calculated. Thus if the results from two dilutions of a test sample, and three concentrations of the standard reference hormone give fiducial limits of 87%–115% it is 95% likely that the true concentration of the hormone lies within these values. Equally if one test fluid gave a value of 100 pg/ml with these fiducial limits, then a value of 90 pg/ml in a second test fluid cannot be taken as showing a drop in the concentration of the hormone. On the other hand a value of 80 pg/ml is probably significantly different from the first test sample (which assayed at 100 pg/ml).

The fiducial limits of assays of ACTH done by the cytochemical segment method varied from 87%–115% to 98%–102% (Daly et al. 1974 b). We took 100 consecutive assays from the notebook of a medical registrar who was working with us and using these assay clinically; values of individual samples ranged from 4 to 320 pg/ml. Each plasma was assayed in duplicate, that is at two dilutions (e. g. 1/100 and 1/1000). It was noteworthy that in 95 of these, the duplicates agreed to $\pm 15\%$ or better than that, only 5 being outside this range. Of those in which the duplicates agreed within the acceptable $\pm 15\%$ limit, 78 had duplicates which agreed within $\pm 10\%$. Thus it appeared that the fiducial limits calculated for a few assays gave an indication of the spread of variability which, in practice, was encountered, with 5% falling outside these limits and most giving reproducibility greater than the outer limits, as would be expected if the reproducibility depended on a Gaussian distribution of values of the mean, so that two-thirds would vary by up to only one standard deviation.

4. Accuracy

The accuracy of a bioassay is best assessed by the percentage recovery of the pure compound, or reference preparation of the hormone, added to the sample before analysis (Borth 1952). In practice, an aliquot of a sample is first assayed for the hormone content of the sample. A known amount of the standard hormone is added to the sample and another aliquot is taken for assay. If the first sample assayed at 10 mU/ml and the second, after adding 100 mU/ml, assayed at 105 mU/ml, the percentage recovery was 95%.

5. Specificity

Related to accuracy and parallelism is the question of specificity. If a recovery experiment produces nearly 100% recovery it is clear that, given a certain concentration of the specific hormone, the assay will be capable of measuring it accurately. But it still does not tell us much about whether the activity measured in a test sample is solely the activity of the hormone we are seeking to measure. To some extent we gain some reassurance by finding that the dose-response of the test sample is parallel to that of the standard preparation of the hormone. But this is not complete proof that the active compound in the test fluid is indeed this hormone. Hence in establishing an assay it is important to test the response, in the selected system, of other compounds or hormones which might mimic the effect of the hormone itself. And the final proof of specificity is often the fact that a specific antibody to the hormone will nullify the biological action of the test fluid in the assay system.

In testing for specificity a powerful tool is the use of known physiological responses. For example, in establishing the first cytochemical bioassay, that of corticotrophin, there could have been some doubt whether or not the unexpectedly low, but measurable, values for the levels of ACTH in certain patients could be due to some biologically active factor, other than ACTH, which was normally present in plasma. However, the plasma of a volunteer, 3 h after being given 100 mg of cortisol to inhibit the secretion of ACTH, recorded only 35 fg/ml of ACTH-like activity (Rees et al. 1973 b). The way the biological activity which was measured by this assay decreased after this intravenous injection, and the low values finally obtained, gave credence to the view that ACTH was the only component of normal plasma to be measured by this assay.

6. Sensitivity

In the context of bioassays this factor is defined by the smallest concentration of the hormone which produces a measurable response significantly above that produced by the vehicle for the hormone itself. Borth (1952) suggested it could be defined as the smallest amount which could be recovered after adding the pure compound to the sample before analysis, i. e. in a recovery experiment. Equally this term is used as a measure of the smallest change in concentration that can be measured by the assay system.

7. Variability

In assessing an assay it is helpful to know the interassay and intra-assay variations. If a single sample is measured on several different occasions, how much do the values vary? To some extent these questions are included in such measurements as the fiducial limits because these limits indicate that, if you were to repeat this assay 100 times, the results of 95 would fall within these limits. However, it may be useful to have information on interassay variability just in case there is a peculiar factor which may cause unforeseen changes in values. It is often worth the extra bother to measure the same sample on different occasions, but before deterioration of the specimen; the results are frequently salutary.

II. Problems of Bioassays

1. Different Potencies Depending on Route of Administration

There is plenty of evidence that the potency of the same preparation of a hormone can be different depending on how it is injected into the test animal, e. g. whether by subcutaneous or intravenous injection. This has led to some confusion in conventional bioassays. The different potencies produced by the various routes of administration reflect problems in the physiology of the hormone, such as whether it mixes with the blood volume or the interstitial water volume, or whether it is metabolized in this or that tissue or organ. To take another example, Lipscomb and Nelson (1962), in describing their bioassay of ACTH, found it was less sensitive if the hormone or test solution was injected into the jugular vein than when it was injected virtually directly into the adrenal gland. The ACTH was apparently ten times less potent by the former route. But the object of a bioassay is not primarily to tell you about the metabolism of the hormone; its main role is to disclose how much of the biological activity, associated with the hormone, is present in the sample. Accordingly Lipscomb and Nelson (1962) used the retrograde injection of ACTH directly up the adrenal vein for their assay and measured the response 90 s later. This question of the potency varying with the route of administration was an unfortunate complication which arose from the nature of in vivo bioassays which could not use the direct application of the hormone to the target tissue in the way achieved by Lipscomb and Nelson.

2. Problem of Desialation

Some authorities (e. g. Ekins 1976) have been concerned that, conceptually, in vitro bioassays should not be able to distinguish between glycoprotein hormones with their intact content of sialic acid, and the desialated molecules which may be equipotent in in vitro studies. The significance of the sialic acid residues is that their presence renders these molecules less susceptible to degradation in vivo. In contrast, removal of sialic acid exposes galactose residues which enhance the uptake, and degradation, of the hormones by other organs in the body, particularly the liver (Tsuruhara et al. 1972). These problems have been discussed by Bangham et al. (1972).

The worry about alleged shortcomings of in vitro bioassays seems divisible into two practical fields of study. The first concerns the unitage of biological activity ascribable to purified preparations of the desialated and intact hormone; the second is how you define activity in a biological fluid.

Before discussing these problems, it will be helpful to look critically at what conventional bioassay actually measures. In vivo bioassays involve injecting the test preparation, or the standard reference preparation, into an animal and measuring a biological effect of the hormone perhaps hours or even days later. The final measurement, therefore, is influenced by a number of physiological factors quite apart from the immediate response of the target cells to being presented with the hormone. It is, in fact, a combination of at least the following components:

1) The mixing volume of the hormone from the initial site of injection.
2) The metabolism of the hormone at sites in the body other than the target cells of that hormone.
3) The immediate effect of the hormone on its target cells.
4) The time course of the hormonal response of the target-cells, e.g. whether it involves either a refractory interval after the initial response or a secondary response (the latter will be discussed in Chap. 15)
5) The time during which the hormone molecules are allowed to act on their target cells (which involves the half-life of the hormone in the circulation).

We may now return to the question of units of biological activity of a desialated hormone. We take a number of laboratory animals, let us say rats, and we inject some with various amounts of the intact hormone and some with equimolar doses of the desialated hormone, and we measure the biological effect at some, possibly extended, time after the injection. We may find that the effect of the desialated hormone is less than that of the intact hormone; let us say it is half the activity of the intact hormone. From this it might be thought that it follows that the potency of the desialated hormone is half that of the intact hormone, on an equimolar basis. This could be used to ascribe a biological potency to the desialated hormone. So far, there would appear to be no difficulty, provided an in vivo assay is used. But someone will then show that the immediate potency of the desialated hormone, tested in an analogous way to the Lipscomb-Nelson assay, i.e. directly on to the target organ with measurement of the immediate response, is equal to that of the intact hormone. As a result, the international unitage ascribed to the hormone becomes discredited. This is comparable to the very real worries of deciding whether one should use the intravenous or the subcutaneous 'units' of activity.

These difficulties occur only because the 'activities' recorded by in vivo assays are composed of two factors, namely the relative potency on the target organ (or immediate activity; 3 above) and the time course over which this potency has been applied (4 and 5 above), modulated by other physiological factors (such as 1 and 2 above). Thus it would be more rigorous to express the relative potencies of hormones by their immediate effect and to include a factor which encompassed the physiological effects. It might be possible to describe the desialated hormone as equipotent to the intact hormone on a molar basis with a factor of 50% when used in vivo. This factor would be determined in normal human subjects rather than in laboratory animals. Equally a hormone could be described as having, for example, 20 international units of activity per mg with a factor of 0.2 if used by intravenous injection or 0.3 if used subcutaneously.

This leads to a consideration of how you define activity in a biological fluid such as plasma. If we inject the fluid into rats we will measure the effect of both intact and desialated hormone. Thus because the time response of most in vivo bioassays is relatively long, we will have less activity if we have more of the desialated molecules present in the fluid, merely because they will be removed more rapidly from the circulation and so will have less time to act on the target tissue. So we will believe that the circulating level of this hormone, in this individual, is lower than it actually is, and such low estimates could be seriously misleading. It is preferable to give the results in terms of the immediate activity of the hormone present (3 above) rather than on the uncertain activity-time effect, particularly since, in this case, the 'time' factor is determined by the physiology of the rat and not of the human subject. If the time factor is important, it should be measured in the human subject itself, as was done in the study on the substituted α^{1-18} ACTH discussed in this chapter, Sect. C.IV. or by measuring the rate of fall in the blood concentration of the hormone after some stimulation.

Consequently to me there is a basic fallacy in the argument that the drawback of in vitro assays is that they do not give information about the time over which a hormone will act in vivo so that they will not distinguish between the desialated and intact hormones. The basic fallacy involves an error in understanding the function of a bioassay. In my view a bioassay should tell you how much biological activity is present in the sample at this particular moment. Physiological studies, to extend the bioassay, may be needed to discover for how long this particular biological activity will be maintained in the subject. And testing this in the rat or guinea-pig, although possibly of use in establishing the unitage of standards, will not give a sure indication of the longevity of action in the human generally, or in a specific human in whom the relevant physiological mechanisms may be altered by the hormonal status related to the condition of that particular human subject. In many ways, to say that in vitro bioassays do not reveal the longevity of a desialated glycoprotein hormone in the circulation is like condemning a pH meter for recording, correctly, that the pH of a fluid is 7.25 ± 0.01 without telling you about its buffering capacity (i. e. that the pH of 10 ml of this fluid will change to pH 7.8 if you add 2 ml of 0.1 *M* alkali whereas the pH of another, at the same initial pH, will be unaffected). It is true that in many conventional in vivo bioassays the relative potencies recorded were clouded to some indeterminant degree, by the time course and other physiological events such as 1, 2 and 5 listed above. My view is that it is better to analyse and record these effects separately.

One of these effects, namely the activity/time course of responses, has already been used in cytochemical bioassays to reveal differences between very similar molecules (as will be seen in this chapter, Sect. C.IV.). For example, although α^{1-39} and α^{1-24} ACTH can both be assayed by the initial response they produce at a certain time, the time courses of the responses they induce in their target cells are sufficiently dissimilar for them to be discriminated (Chayen et al. 1976; also see Chap. 15). It is possible that similar differences may be found between the intact and desialated forms of glycoprotein hormones but this has not yet been investigated. Even if this was not found, and if it were necessary to assay the desialated hormone separately from the intact hormone, it might be possible to use a two-stage assay. In this (cf. the assay of thyrotrophin-releasing hormone in Chap. 14) the hormone might first be exposed to liver and then be transferred to the target organ.

3. Secondary Effects of Hormones

A different, and possibly more relevant, example of physiological effects in the whole animal influencing the biological performance of a hormone may be the in vivo production of somatomedins by growth hormone. It seems clear that the full biological potency of growth hormone cannot be measured by an in vitro study of its influence on cartilage. In contrast, in experiments on whole animals, the effect of somatomedins produced, for example, through the action of growth hormone on the liver, would contribute to the biological activity of the growth hormone used in such studies. This, however, is not an insurmountable problem; it is somewhat related to the measurement of the activity of hypothalamic regulatory factors. For example, thyrotrophin-releasing hormone (TRH) can be assayed by the in vitro cytochemical bioassay system in a two-stage procedure (Gilbert et al. 1975). The TRH is applied to segments of pituitary gland causing them to secrete thyrotrophin (TSH) in a dose-related fashion; the TSH is measured by its effect on thyroid follicle cells. There is no obvious reason why segments of liver should not be used to test how the liver has modified the effect of growth hormone, or influenced desialated gonadotrophic hormones; the culture medium would then be applied to the target tissue very much as the culture medium in which the pituitary segments were treated with TRH is used as a source of TSH. However, it should be emphasized that there is no reason for believing that the liver of whatever animal is used in the two-stage assay will correspond in its behaviour to the liver of a particular patient, which already may be influenced by the peculiar hormonal status of that patient. As in the case of desialated hormones, we are here entering the realm of human physiology and leaving what can be expected of a bioassay system. On the other hand, it is true that the cytochemical bioassay system can be used to test the effective performance of critical tissues in the patient. For example, suppose that the circulating levels of biologically active gastrin in a particular patient are unrelated to the secretion of gastric acid. If a biopsy sample can be obtained from the gastric fundus it will be possible to test how responsive these parietal cells are to standard concentrations of synthetic gastrin. The fault may lie in target-cell response rather than in the circulating level of the hormone. Equally the discrepancy could be due to antibodies directed either against the parietal cells or against gastrin. These can be discriminated by slight variants of the normal cytochemical bioassay procedure. Thus the in vitro cytochemical bioassays can help in elucidating some aspects of endocrine physiology, but this is not the primary concern of bioassay. This is to measure how much of a particular hormonal activity is present in the test fluid at the moment of assay.

B. Cytochemical Bioassays

I. Basic Approach

1. Biochemical Effect of a Hormone

The fundamental concept underlying the cytochemical bioassays is very simple: when a hormone acts on its target cells it produces changes in biochemical systems which allow those cells to produce the physiological effect of the hormone. Thus when gastrin acts on parietal cells it causes, or helps to cause, acid secretion. The hormonal activity is known as 'gastrin'; the target cells are the parietal cells of the gastric fundus; the physiological effect is the secretion of acid. Yet to stimulate these cells to secrete acid, the hormone must have caused some biochemical change in those cells. The most notable biochemical change which has been recognized at present is an elevation of activity of the enzyme, carbonic anhydrase. This seems reasonable in the light of our present understanding of gastric acid secretion.

Table 3.1. Cytochemical bioassays currently in use (for details please see relevant chapter)

Biological activity	Target cell	Biochemical activity affected	Function	Sensitivity
ACTH	Adrenal zona reticularis	Ascorbate depletion	Steroido-genesis	5 fg/ml (c.a. 5×10^{-4} μU/ml)
TSH	Thyroid follicle cell	Lysosomal labilization	Production of thyroid hormones	4×10^{-5} μU/ml
LATS	Thyroid follicle cell	Lysosomal labilization	Production of thyroid hormones	–
LH	Corpus luteum	Ascorbate depletion	Steroido-genesis	10^{-4} mU/ml
Gastrin	Gastric parietal cells	Carbonic anhydrase	Acid secretion	5 fg/ml
PTH	Distal convoluted tubule	Glucose-6-phosphate dehydrogenase	Movement of calcium	5 fg/ml
TRH	Pituitary	Production of TSH[a]		–
CRF	Pituitary	Production of ACTH[a]		–

[a] Measured by the appropriate cytochemical bioassay.

Consequently it is appropriate to measure gastrin activity by its ability to increase the activity of carbonic anhydrase in the parietal cells. And the fact that cytochemistry is now a precise form of cellular biochemistry (as discussed in Chap. 1) means that we can now measure such changes just in the target cells. The same general concept underlies all the cytochemical bioassays produced up to the present time (Table 3.1).

The selection of a biochemical mechanism which is involved in the physiological disclosure of the hormonal activity gives greater specificity to these assays. It is likely that each polypeptide hormone acts by binding to a specific site an the plasma membrane of the target cell and that this interaction stimulates adenyl cyclase or, in some instances, guanyl cyclase. It might have been thought adequate to measure the activity of these enzymes. However, other substances which cause perturbation of the plasma membrane may also activate the relevant cyclase. Greater specificity therefore is obtained by using both the binding to the target cell and the specific biochemical response within that cell induced by such binding.

2. Hormonal Status of Tissues

In conventional bioassays of hormones influenced by, or derived from, the pituitary gland, it is normally necessary to use hypophysectomized animals in order to obtain a sufficiently sensitive and invariable response to exogenous hormone. In the cytochemical bioassays this is replaced by removing the target organ from the hormonal influence of the animal and maintaining it, in non-proliferative adult organ maintenance culture. For all the hormones studied up to the present time, 5 h have been sufficient to reduce the hormonal influence of the animal to a sufficiently low level. This can be tested quite simply. If a particular biochemical activity is under hormonal influence, that activity will decline to a basal level when the tissue is removed from the animal and maintained in a maintenance culture. For example, if we examine the glucose-6-phosphate dehydrogenase activity in the distal tubules of a sample of the kidney of a guinea-pig removed at death, the activity is high but is greatly diminished in segments of the same kidney at the end to the 5-h culture period. This is the type of response expected of a hormonally controlled biochemical activity, provided that the culture conditions are not deleterious. Since the histological preservation of the tissue appeared to be acceptable and the activity of other enzymes, such as succinate dehydrogenase, was not diminished, it seemed likely that glucose-6-phosphate dehydrogenase activity in these cells was indeed under the influence of some circulating factor. The activty recovered when parathyroid hormone was added to the culture medium, indicating that the circulating factor could be parathyroid hormone (Chambers et al. 1976).

3. Physiological State of the Target Cells

The use of Trowell's (1959) maintenance culture in cytochemical bioassays is discussed in Chap. 4. In that chapter (Sect. 4.B.2) is recorded the early work which showed that cells in segments of tissue show peculiar changes due to the trauma of excision from the animal and being placed in the in vitro culture conditions. The cells recovered fully 5 h after being placed in the culture system. It was therefore likely that segments, used in the cytochemical bioassay systems, would respond with greater sensitivity if they were allowed to recover from the trauma of excision before attempting to stimulate them with the selected hormone. In studying the effect of ACTH on segments of guinea-pig adrenal gland it was found that they were virtually non-responsive after 3 h in maintenance culture but fully responsive at 5 h. Similar results were found for thyroid segments; furthermore they were no more responsive after overnight culture than they were after 5 h (Ealey 1979).

4. Need for Essential Metabolites

Another advantage of the 5-h culture period is that it can be used to allow the cells to accumulate any essential metabolites or to become 'primed' with essential hormones. This effect can be seen in our studies on the adrenal gland (Table 3.2). Segments were taken from the two adrenal glands of each of four guinea-pigs; one was chilled immediately after removal from the animal, and the others after 5 h in Trowell's T8 medium either alone or reinforced with one of two concentrations of ascorbate. The pH was maintained the same, pH 7.6, for all specimens.

Table 3.2. Influence of ascorbate on the activity of NADP-dependent oxidative activity in the adrenal zona reticularis in vitro (Chayen et al. 1976)

	Activity (relative absorption/unit area/unit time) in:			
Dehydrogenase	Biopsy	T8 medium	T8 + 10^{-4} *M* ascorbate	T8 + 10^{-3} *M* ascorbate
Glucose-6-phosphate	307	118	185	291
	282	189	–	271
	190	146	–	190
	437	261	369	458
6-phosphogluconate	295	205	–	290

Firstly it was obvious that the levels of activity of glucose-6-phosphate dehydrogenase varied considerably between animals, and that it dropped markedly after 5 h in the normal culture medium. Ascorbate, at 10^{-3} *M*, regenerated the activity to about the level found in the biopsy. Similar results were found with the second dehydrogenase enzyme of the hexose monophosphate pathway, namely 6-phosphogluconate dehydrogenase (Chayen et al. 1976).

Both these enzymes depend on sulphydryl groups for their full activity. So it was not surprising that the levels of ascorbate, or the reducing potency, in the cells of the zona reticularis also showed similar fluctuations. In a particular study (Table 3.3) even though the concentration of ascorbate was much lower in the cells of the zona reticularis of the first animal, due possibly to different degrees of stress at death releasing different amounts of ACTH, the 5-h culture in the presence of a sufficient concentration of ascorbate resulted in the cells of both animals having comparable reducing potency. It is noteworthy that the cells of the zona fasciculata apparently responded neither to the trauma of death nor to the ascorbate in the medium (also see Chayen et al. 1972a)

Thus it was possible to replenish the stores of ascorbate during the 5-h culture period. This has considerable practical advantage in that the sensitivity of an assay based on the depletion of ascorbate obviously depends on the concentration of this substance present prior to challenge by the hormone.

Other tissues, and their response to their specific hormone, may require other metabolites or even other hormones to prepare the cells for maximal response to the specific hormone being tested.

5. Summary of the Basic Features of Cytochemical Bioassays

1) They are all within-animal assays, utilizing pieces (segments) of the target organ from the same animal and so eliminating the considerable between-animal variability that has contributed to the imprecision of conventional bioassays. Where the organ is paired, as is the adrenal gland, it is necessary to show that the left and right organs have identical biochemical activities and responses before segments of both are used in a single assay. This identity of response has been shown for the right and left adrenal glands and the right and left lobes of the thyroid gland of the guinea-pig.
2) The segments are maintained in non-proliferative organ maintenance culture for 5 h to allow the cells to recover from the trauma of excision and, where necessary, to replenish essential metabolites. This process allows the cells to become free of

Table 3.3. The effect of ascorbate on the reducing potency (relative absorption (E_{640})/unit area/unit time) in cells of the adrenal cortex maintained in vitro

Animal	Medium	Zona fasciculata	Zona reticularis
1	T8	173	68
	T8 + ascorbate (10^{-3} *M*)	167	140
2	T8	136	128
	T8 + ascorbate (10^{-3} *M*)	143	150

the hormonal influence of the animal and therefore permits hormonally controlled biochemical activities to achieve a basal, unstimulated level.

Increased sensitivity is provided by the recovery period and by the provision of essential metabolites, where required. It is also aided by the method of maintenance culture in which the tissue is not immersed in the medium (as discussed in Chap. 4).

It is very likely that increased sensitivity also is obtained by the fact that the hormone will be applied to an intact tissue or, in the section assays, to an intact section. That is to say, there is some reason (see Chap. 4) to believe that the response of cells to a hormone depends not only on the response of the cells to hormone molecules binding directly to them but may also be influenced, or even be induced, by the binding of the hormone to adjacent cells possibly by some form of transmembrane potential. This is an additional reason for not using an isolated-cell preparation, as discussed in Chap. 4 (Sect. A.II).

3) At the end of the 5-h culture it may be necessary to replace the medium with fresh medium to remove the presence of metabolites which may impair the sensitivity of the response of the cells to the hormone. For example, segments of the adrenal gland produce considerable quantities of corticosteroid which can influence the response of the cells of the zona reticularis to ACTH (Chayen et al. 1974 b). In most assays, however, it has not been found necessary to change the medium and allow the tissue some time in the fresh medium before exposing it to the hormone.
4) Normally, after the 5-h culture period the medium is replaced by fresh medium, including any additives that were found necessary, together with one of four graded concentrations of a standard preparation of the hormone, or a suitable dilution of the plasma (usually 1:100 or 1:1000) in which the hormonal activity is to be assayed. The period of exposure to the hormone is usually a matter of only minutes, in order to measure the initial response. Part of the sensitivity of these assays is due to the fact that they measure the initial velocity of response and not the time-dependent accumulation of an end product (Chayen et al. 1974 b).
5) At the appropriate time the hormone response is stopped by chilling the tissue to -70° C. The biochemical change induced by the hormone is measured in sections, in which a coloured reaction product is produced cytochemically (as discussed in Chap. 1 and 5). The amount of reaction product formed specifically in the target cells is measured by scanning and integrating microdensitometry (see Chap. 6). A great part of the sensitivity of these assays is due to the development of cytochemistry and particularly to the sensitivity of the microdensitometer, as is explained in Chap. 6.

II. Cytochemical Section Assays

1. Basic Procedure

In the cytochemical section assays the procedure is the same as in the segment assays in that segments of the target tissue are maintained in vitro for 5 h. In some assays the segments are then primed with the hormone to be assayed; the concentration of hormone applied is at least one order less than the lowest concentration that has a detectable effect in the cytochemical test used for measuring the biochemical change induced by the hormone. Thus, for example, the lowest concentration of ACTH that can be detected in the ascorbate-depletion procedure (Chap. 7) is 5 fg/ml and this can be assayed by its effect at 4 min. Consequently in the section assay the segments are primed by being exposed for 4 min to fresh T8-ascorbate medium containing 0.5 fg/ml of ACTH. The primed segments are then chilled to −70° C and are sectioned at sufficient thickness to encompass a full cell. Thus for the cells of the zona reticularis of the guinea-pig the sections have to be 20 μm thick; for the rather flatter thyroid follicle cells, in the assay of TSH, they may be 12 μm thick. The chilled tissue must be used within 2 or 3 days of being chilled. The ability of the cells to respond to hormone decreases rapidly with longer storage at −70° C.

The sections are then exposed either to one of four graded concentrations of the standard preparation of the hormone or to the plasma at a suitable dilution. Again, as in the segment assays, this is usually at 1:100 or 1:1000 dilution of the plasma. The maximum initial response to the hormone in sections is much more rapid than in segments, being achieved in tens of seconds. Thus the time for the maximum initial response of cells of the zona reticularis in sections of guinea-pig adrenal gland is only 60 s, as against 4 min when the cells are in segments of the gland.

The same cytochemical and microdensitometric procedures are used on the sections subjected to the section assays as on sections cut from the segments in the segment assays.

The cytochemical section assays were developed to accommodate the need for many more samples to be assayed in a normal working week than could be done by the segment assays. The former had been instituted as a research procedure but their success, and the recognition of the importance of bioassay even in the era of radio-immunoassay, brought the demand for such assays which could be used routinely. The problem was this: for the assays to be within-animal assays all the tissue had to be derived from one animal. Because it was not possible to gauge, with any degreee of accuracy, what the hormone level was likely to be in the test sample, and because we required each sample to be tested at two dilutions, the calibration graph had to cover four points (usually covering four orders of concentration). This required four samples of the tissue. It is frequently not possible to obtain more than six segments of an organ; this applies especially to the thyroid or adrenal glands. Seeing that four segments are used for the calibration graph, this leaves only two for the test sample, diluted let us say 1:100 and 1:1000. It follows that each sample required one guinea-pig. Each required a separate calibration graph. The use of sections, in place of segments, meant that the same calibration graph served for all samples, and very many more samples could be assayed, each at two dilutions. The restriction in the section assays is now mainly the speed, and tedium, of making the measurements, and the dimensions of the racks used for holding the sections. These section assays also make it possible to include a quality control, namely a sample containing a known concentration of the hormone.

I am told that in one laboratory, over a certain period, two workers using the same cryostat and microdensitometer (but at mutually convenient times) have each produced 40 ACTH assays, each in duplicate, each week. A total of 30 by 1 worker in a week does not seem unduly arduous.

2. Critical Importance of Colloid Stabilizers

As will be explained in Chap. 5, fresh sections of undenatured protoplasm lose about 50% of their total nitrogenous content if immersed in a buffer at a pH of between 7 and 8. For many hormones to be at their most effective, the sections must be exposed to hormone at about pH 7.6. As discussed in Chap. 5, all the nitrogenous matter and so-called soluble enzymes can be retained inside the sections, even at this pH, provided that the sections are protected by a sufficient concentration of a colloid stabilizer in the reaction medium. However, when sections are fully stabilized in this way, they are so stable that they will not respond to stimulation by the hormone. The development of most section assays therefore involves a painstaking exercise in balancing the need to protect the cells on the one hand, as against the requirement that they should still be capable of responding to the hormone. In the section assay for TSH, where it was essential not to over-stabilize the lysosomal membranes, the effect of a very low concentration of a stabilizer has been enhanced by the use of acetate ions, which in other studies were shown to cause no change, over a few minutes, in the lysosomal membranes of thyroid follicle cells.

It is, in fact, highly significant that the various processes used in preparing and exposing the sections to the hormone have not impaired the sensitivity with which the cells respond. These processes include chilling the tissue, sectioning it at -25° C, flash-drying the sections and then rehydrating them in the presence of the stabilizer. The unimpaired response of the cells is probably the best validation of the cytochemical procedures for preparing sections.

C. Special Features of Cytochemical Bioassays

I. Details of the Assays

1. Dose-Response Calibration Graphs

For most of the assays the dose-response graphs show a linear relationship between the logarithm, to the base 10, of the concentration of the hormone applied and the extent of the cytochemical response induced by the various concentrations of the hormone (as in Fig. 3.3). As discussed previously (this chapter Sect. A.I.3.) the index of precision (λ) is given by the deviation of all the points from the calculated regression line (s, or s_{yx} in some notations) divided by the slope (b). Consequently it is advantageous to have a steep slope so that b will be large and λ will be correspondingly small. However, for most of the cytochemical bioassays the slope is rather shallow. This is compensated by the precision of measurement so that s_{yx} is also small. In the cytochemical bioassays this factor, s_{yx}, can be measured by using two serial sections for each point and plotting the mean value from each section on to the graph. It would be more rigorous, but very time consuming (since the measurement is the most exacting part of these assays), to use five sections for each point on the graph and to calculate the standard deviation, or standard error of the mean. When this has been done the results have not been materially different from using the results from two serial sections or even the mean of the readings in the two sections.

In general, for estimating s_{yx} we take the mean value of the means of the readings in each slide. We know that when we measure the response in five sections from each segment of tissue treated with its particular concentration of the hormone, the results agree to about $\pm 5\%$ (e.g. Chayen et al. 1972a for ACTH; Bitensky et al. 1974a for TSH; Loveridge et al. 1974 for gastrin).

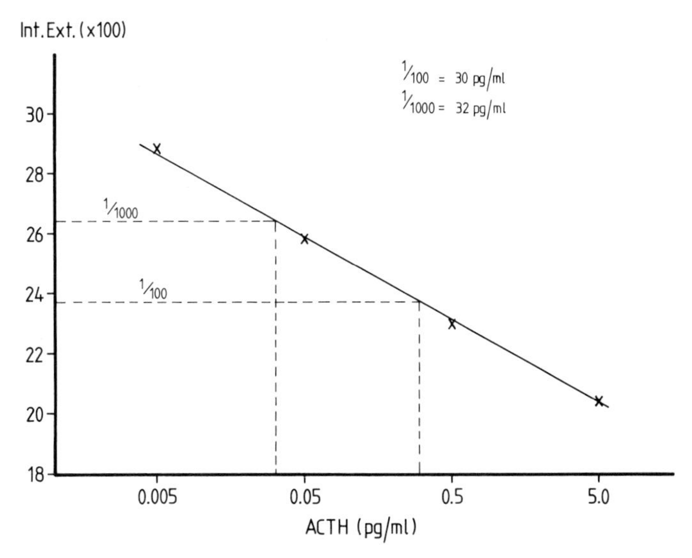

Fig. 3.3. Calibration graph (*crosses*) of segment assay of ACTH with two dilutions of a plasma (1:100 and 1:1000). The reducing potency (ascorbate concentration) remaining in the cells of the zona reticularis in sections of these segments, after exposure to the appropriate concentration of the hormone, is measured (Int. Ext. × 100) and plotted against the concentration of ACTH that has been applied. Although the slope (b) is shallow, the deviation of the readings from the regression line (s_{yx}) is small, so giving an acceptable index of precision. The absorption (Int. Ext. × 100) in comparable cells in sections of the segments exposed to each dilution of the plasma is read off this calibration regression line (*broken lines*). When corrected for dilution the values were in good agreement (30 and 32 pg/ml), indicating good parellelism of response

The index of precision for cytochemical bioassays is usually about 0.1; occasionally it is below this value, as will be seen when each assay is discussed individually. Following Cornfield (1970), a λ of 0.1 means, in operational terms, that the log dose estimated from a single observation has a standard error of 0.1, since λ is, in practice, the standard error of the log dose. Thus to convert this to the error of the estimated dose, we take the antilog of 0.1, which is 1.26. Consequently a single measurement can be 1.26 times (or equivalently less) the measured value. This is confirmed by the fiducial limits of these assays, where they have been calculated, that is in assays where five sections were taken for each point. For example, the fiducial limits of the segment assay for ACTH ranged from 87%–115% down to 98% –102% (Daly et al. 1974b).

2. Parallelism

The significance of parallelism in bioassays generally has already been considered (this chapter Sect. A.I.2.) and some of the more significant features of lack of parallelism will be discussed later (this chapter Sect. C.II.). Parallel responses are of especial importance in the cytochemical bioassays because the design of these assays requires that each sample is assayed at two dilutions (as in Fig. 3.3). Provided that like is being compared with like, the responses should be parallel. Consequently the results of each dilution, when corrected for dilution, should agree to ±15% or better than this. If they do not, the failure may be due either to experimental error, which can be tested by repeating the assay, or to a true lack of parallel response, as will be discussed later (this chapter Sect. C.II.). Thus the two samples, at different dilutions, give a check on the precision of the particular assay being done. If the λ value for this assay is 0.1 (i. e. the error of the estimated dose should be 1.26), the results obtained from the two dilutions should agree to within $\pm 26\%$; if they do not, there must be either experimental error, or there is some 'unlike' factor in the sample.

This criterion of parallelism can be applied even when the dose-response graph is not linear, as in the cytochemical segment assay of TSH (Bitensky et al. 1974 a).

II. Deviations from Parallelism

Apart from experimental error, deviations from parallelism in the responses at two dilutions require investigation. Frequently such investigation discloses facts of considerable importance. As far as our present experience goes, deviations from parallelism can be indicative of one of the following phenomena:

1) The sample may contain an active moiety, other than the hormone used as the standard preparation, which closely simulates the biological activity of the hormone but produces results which are not parallel to those of the standard hormone preparation (Fig. 3.4 a). This possibility can be examined by seeing if the biological activity is neutralized by an antibody specific to the hormone.
2) The sample may contain matter which alters the time course of the response of the hormone. For example, Loveridge et al. (1978) found that acetyl choline facilitated the response of carbonic anhydrase, in parietal cells, to gastrin. This phenomenon can be tested by examining the time course of the response. Indeed it is often worth measuring the time course when peculiar responses occur.
3) The sample may contain antibodies directed against the hormone. The antibody-hormone complex may be retained at relatively low dilutions (e. g. 1/100) so that the hormone is not allowed to express its full activity. At greater dilutions (such as 1/1000 or even 1/10,000) the antibody-hormone complex can dissociate. Thus it may be found that the hormone content of three dilutions of the sample (1:100, 1:1000; 1:10,000), corrected for dilution, were 5, 25 and 100 pg/ml (or μU/ml); this could be due to the dissociation of the antibody-hormone complex partially at 1/1000 but completely at 1/10,000 dilution. If an excess of the antibody is present in the sample, then this type of investigation is best done by a recovery experiment, i. e. by adding a known amount of a standard preparation of the hormone to the sample. This allows greater dilution of the sample than might be permitted by the concentration of the hormone endogenous to the sample.
4) The sample may contain antibodies directed against the target cells, or more specifically against receptors in or on the target cells. Consequently it will be found that, in low dilutions, the sample will produce very little biological response even if a considerable concentration of a standard preparation of the hormone is added to the initial sample from which the dilutions are made. The presence of antibodies of this type can be investigated by maintaining the target tissue, in segments, for the 5-h maintenance in vitro, in the presence of the sample (at a dilution of 1:10 or 1:100 in the normal culture medium). The culture medium is then replaced with the normal medium containing one of a graded range of concentrations of the standard preparation of the hormone. If antibodies are present in the sample, they will have attached to the cells during the 5-h

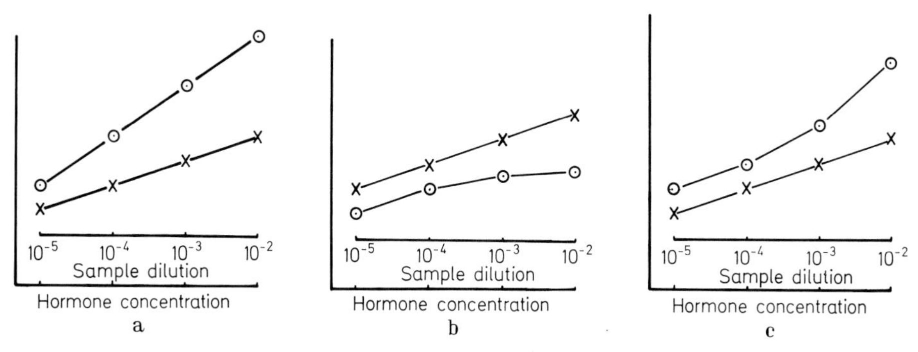

Fig. 3.4a–c. Hypothetical results to show types of non-parallel response. (*Crosses* indicate the calibration graph measured against a standard preparation of the hormone; *circles* show the hypothetical response of dilutions of the various samples.) **a** Presence of an active moiety which simulates the activity of the hormone. **b** An antagonist to the hormone or to the receptor causes inhibition at higher concentrations but becomes diluted-out at higher dilution, allowing parallelism at dilutions of 10^4 and 10^5. (But it must be noted that a similar response graph may be obtained if the hormone concentration in the plasma sample is too concentrated; it must be diluted until it fits the calibration graph.) **c** At low dilutions an activator may produce greater activity than does the hormone itself but its effect is likely to become diluted-out more rapidly than the influence of the hormone

maintenance period so that these cells will fail to respond, or will respond abnormally, when exposed to the hormone in culture medium even though the antibodies are not present in the medium in which the hormone is presented to the cells.

5) In the possibilities discussed under (3) and (4), the lower dilutions of the sample will show less biological activity, the full activity becoming apparent only as the dilution becomes so great that the effect of the antibody or other antagonist is diluted to undetectable levels (Fig. 3.4 b). An alternative situation is where the sample contains some other biological activity which, at relatively high concentration, simulates that of the hormone. For example, certain prostaglandins can activate the same cellular mechanisms that are the normal province of a hormone. To exert their influence such activating molecules normally have to be present at much higher molar concentrations than are required by the equivalent hormone. If they are present, therefore, their influence is likely to be seen at relatively low dilutions of the sample; at higher dilutions only the hormone itself will have appreciable activity. This phenomenon will make itself apparent by a lack of parallelism in which the lower dilutions have greater activity than expected (Fig. 3.4 c). When such a sample is treated with an antibody specific to the hormone, an appreciable part of the activity (at low dilutions) will not be neutralized.

III. Use of Plasma

At the present time most samples of blood taken for hormone assay are taken for radio-immunoassay. Consequently the blood frequently is allowed to clot and, at a time convenient for the busy clinician, the serum is separated and sent for assay. The time between taking the blood and separating the serum appears to be very variable and of no particular consequence. This procedure may be satisfactory for many radio-immunoassays in that the assay is preferably done on serum and the longevity of the immunoreactive molecules seems to be extensive. It is to be eschewed by bioassayists generally, but especially by those using cytochemical bioassays. Indeed it is one of the advantages of these bioassays that, unlike those using isolated cells, they can assay hormones in plasma; the hormone does not have to be extracted into an artificial medium for presentation to the target cells. The reasons for preferring plasma for use in cytochemical bioassays are as follows:

1. Loss of Biological Activity

The biological half-life (or half-time) of many polypeptide hormones is a matter of minutes rather than of hours. In the circulation, in man, the half-time of exogenous α^{1-24} ACTH was about 9 min (Daly et al. 1974 b); that of endogenous ACTH was 10.4 min (Rees et al. 1973 b). Holdaway et al. (1974 b) left plasma samples at room temperature for 1 h and recorded a decrease in the biological activity of endogenous ACTH of nearly 90%; i. e. the activity after 1 h was about 10% of what it had been when freshly prepared. Ealey (1979) found the half-time of TSH in the circulation was about 20 min, this figure being based on the changes in concentration of the biologically active TSH after it had reached a peak in subjects who had been injected with TRH.

Of especial interest are the results recorded for gastrin-like activity. Hoile and Loveridge (1977) found a marked rise in the circulating levels of gastrin-like activity in subjects given an Oxo test-meal, which is expected to stimulate the secretion of gastrin. The levels rose from 9 to 180 pg/ml 5 min after the 'meal'. (In a second subject the values were 14 to 289 pg/ml respectively.) Such a rapid and marked rise, which was expected on physiological grounds, had not been found by expert radio-immunoassayists who, of course, used serum in their assays. However, more recently, Christofides et al. (1978) appear to have confirmed this rapid and striking rise in gastrin, measured by radio-immunoassay but in plasma, not serum. Why radio-immunoassay could not detect this increase in serum is not immediately obvious.

2. Active Substances in Blood Cells

Radio-immunoassay is unlikely to be influenced by materials which may leach out of blood-borne cells or platelets during the coagulation required for the separation of serum. However, such cells may contain relatively large amounts of biologically active molecules which might activate the biochemical systems used in the cytochemical bioassays. For example, blood contains about 10^{-6} *M* spermidine, all of which is present in the cells (Documenta Geigy, 1970). It has been shown that spermidine at concentrations of 10^{-10}–10^{-6} *M* can mimic the effect of TSH in labilizing the lysosomal membranes of thyroid follicle cells (Gilbert et al. 1977b). This effect may have some relevance to the mode of action of TSH, as will be discussed in Chap. 10. In the present context it may be noted that if all the spermidine were to be released from the cells in the blood during the clotting process, there could be up to 10^{-6} *M* spermidine in the serum which, at a dilution of 1:100 (as used in the cytochemical bioassay) could provide an unwanted source of stimulation.

3. Configuration of Polypeptide Hormones

At present too little is known of whether the configuration of polypeptide hormones is totally stable or whether it may be influenced by the chemical composition of the medium in which they are dissolved. In our own work we found that the standard preparation of PTH (bovine) was very unstable in the normal vehicle so that it appeared to lose biological activity very readily and unpredictably, whereas the PTH-like activity of a sample of plasma, used as a quality control, seemed very stable. Stability was restored to the standard preparation when it was dissolved in plasma taken from a hypoparathyroid subject (D. J. Chambers, personal communication). Indeed to achieve complete parallelism in the cytochemical bioassay of this hormone it has been necessary to add PTH-deficient plasma to the solutions of the standard preparation used for the calibration graph. For this and other reasons it seems possible that, for some hormones at least, plasma may be necessary to endow the hormone molecules with the most potent and stable configuration for their activity.

IV. Physiological Studies

Previously (this chapter, Sects. A.II.2 and C.II.2) it was argued that the object of bioassay is to record the concentration of the particular biological activity by which the hormone is known. It is not, in the first place, the task of bioassay to give physiological information as to whether or not the hormone is of the long-lived variety. On the other hand, it was conceded that the cytochemical bioassays are capable of being used to some effect in such physiological studies. Most of these will be discussed in the context of the relevant hormone. Here it may be helpful to choose a particular example which, in some ways, is akin to the problem of the desialated glycoprotein hormones.

The study involved a substituted α^{1-18} ACTH, with *D*-serine at the 1-position, lysine substituted for arginine at the 17-position and lysine amide at the 18-position. It was known (Keenan et al. 1971) that this molecule had a relatively long half-time in the circulation and therefore was a more potent steroidogenic agent, in life, than the normal α^{1-39} ACTH or the synthetic α^{1-24} ACTH [tetracosactrin (Synacthen)].

In the cytochemical bioassay the response graph was not entirely parallel to that of the Third International Working Standard (Fig. 3.5). The results were sufficiently parallel over the two lower dose-ranges for a relative potency (2 + 2 potency test) to be made. In two separate investigations the relative potency, on a weight basis, was 1.03 and 1.09. Thus on a molar basis the substituted 1–18 peptide was up to twice as potent as the α^{1-39} ACTH. The time courses of the response (Fig. 3.6) showed that the two substances were not exactly identical, the 1–18 peptide producing its maximal effect more slowly, namely at just over 5 min. Thus poor parallelism and a displaced time course indicated that the two molecules were dissimilar. These results could not explain the greater in vivo potency of the substituted α^{1-18} molecule; for this it was necessary to do a more physiological study. A typical procedure for this study was as follows: a volunteer took 2 mg of dexamethasone orally at midnight, and another 2 mg at 08.00 hours on the morning of the test to inhibit the endogenous secretion of ACTH. In the first test 800 ng of tetracosactrin

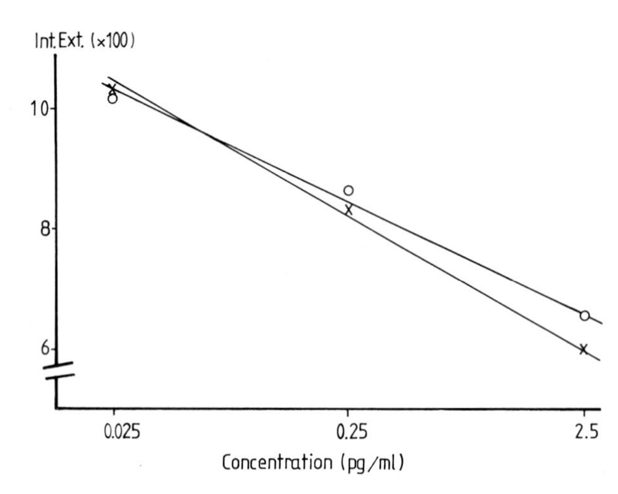

Fig. 3.5. Dose-response graphs of the α^{1-39} (*crosses*) and α^{1-18} (*circles*) ACTH. At the lower concentrations the degree of parallelism was sufficient to allow comparison of relative potencies

were given intravenously; in the second test more than 2 weeks later, 500 ng of the substituted α^{1-18} ACTH were given by the same route. The injections were at 09.00 hours when the morning peak of ACTH secretion, even in unsuppressed subjects, would be at its lowest. Samples of plasma were taken for determining the levels of ACTH (by the cytochemical segment bioassay) and of corticosteroids (by the fluorimetric method of Spencer-Peet et al. 1964).

The results of injecting α^{1-24} ACTH showed that the maximum circulating concentration of the hormone was achieved 4 min after injection; the concentration achieved indicated that the hormone had gone into a volume of about 3–4 litres, compatible with the view that, at this stage, the mixing volume of the hormone was that of the plasma of this female subject alone. The rise in corticosteroid levels in the

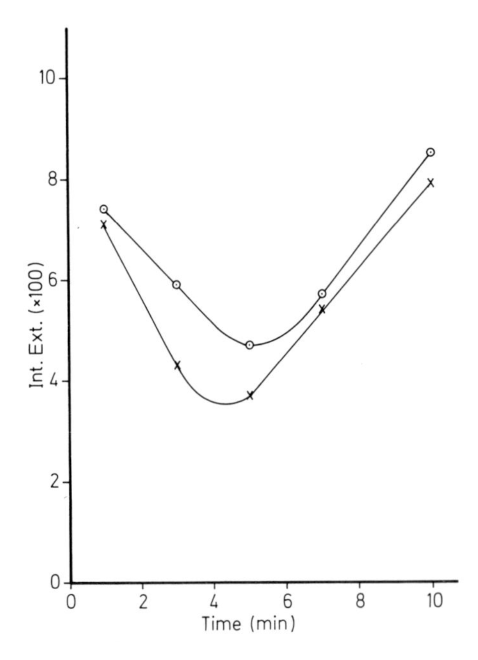

Fig. 3.6. Time courses of the action of the α^{1-39} ACTH (*crosses*) and of the α^{1-18} ACTH (*circles*). The rate at which these act in decreasing the ascorbate concentration (Int. Ext. × 100) in the cells of the zona reticularis was not identical; the two molecules can be discriminated by this phenomenon. However, since both act maximally at about 4 min, they can be assayed at this time, although the different time courses may account for the lack of parallelism shown in Fig. 3.5

blood followed soon after the rise of ACTH levels and declined with them (Fig. 3.7). In contrast (Fig. 3.8) the 1–18 peptide took longer to reach its maximum concentration in the circulation (8 min in one volunteer, 12 min in another and

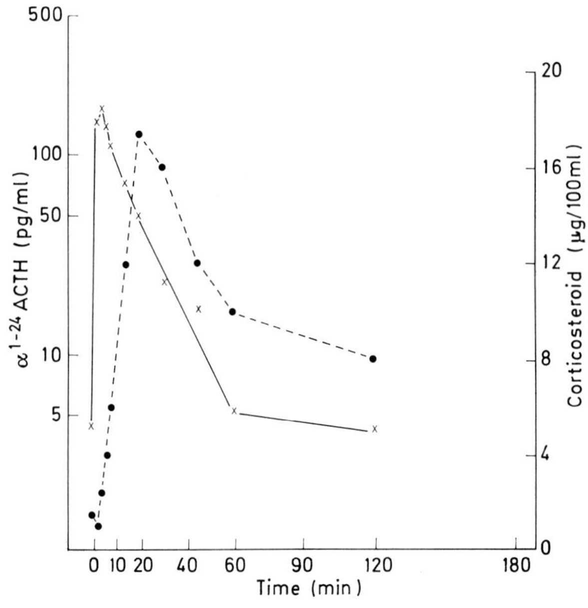

Fig. 3.7. Levels of ACTH (*solid line;* log scale) and of corticosteroid (*broken line*) in the plasma of a healthy, dexamethasone-suppressed subject injected intravenously at time 0 with 800 ng of α_{1-24} ACTH. (Daly et al. 1974 a, p. 337)

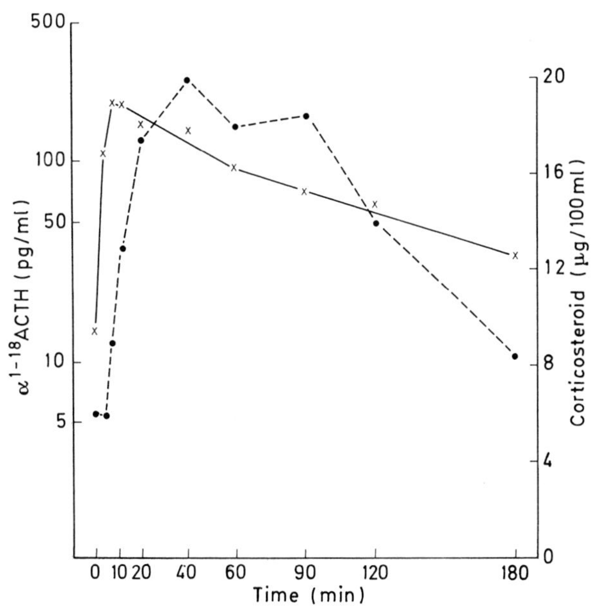

Fig. 3.8. As Fig. 3.7 but following the intravenous injection of 500 ng of the substituted α_{1-18} ACTH. The prolonged retention in the circulation of the substituted ACTH (*solid line*) and its extended effect on the level of corticosteroid in the plasma (*broken line*) are apparent. (Daly et al. 1974 a, p. 338)

15 min after an injection of only 60 ng) and the concentration did not fall rapidly as it did when the 1–24 peptide was injected. (Obviously, the 1–18 substituted peptide was used as the standard for the calibration graph for test samples containing this peptide.) The half-time of the 1–18 substituted form of ACTH was between 51 and 53 min, in different experiments, as against about 9 min for the α^{1-24} ACTH. The production of corticosteroid was equivalently prolonged. Even when only 60 ng of the substituted 1–18 peptide were injected, so that the physiological mechanisms were subjected to physiological concentrations of the hormone-like material, the same pattern was observed (Fig. 3.9). Thus the greater potency of the 1–18 substituted ACTH-like material, in life, was due to its considerably longer half-time in the circulation so allowing more molecules to exert their steroidogenic stimulation over a much longer time period.

Thus the bioassay did all that, in my opinion, a bioassay should be asked to do: it showed that the 1–18 peptide had ACTH-like properties but was dissimilar from the full 1–39 peptide; over the range that the responses were reasonably parallel it measured the concentration of the 1–18 peptide and its relative potency as regards the immediate effect of the hormone-like molecule. As an assay it could not explain the greater in vivo potency of the 1–18 substituted peptide. However, used in conjunction with a simple physiological investigation it could demonstrate the longer half-time of the 1–18 molecule and so explain its greater in vivo potency.

Other examples of how the cytochemical bioassay system can discriminate between different molecular forms which have similar biological activities will be discussed in the relevant chapters dealing with different forms of ACTH and of thyroid stimulating substances.

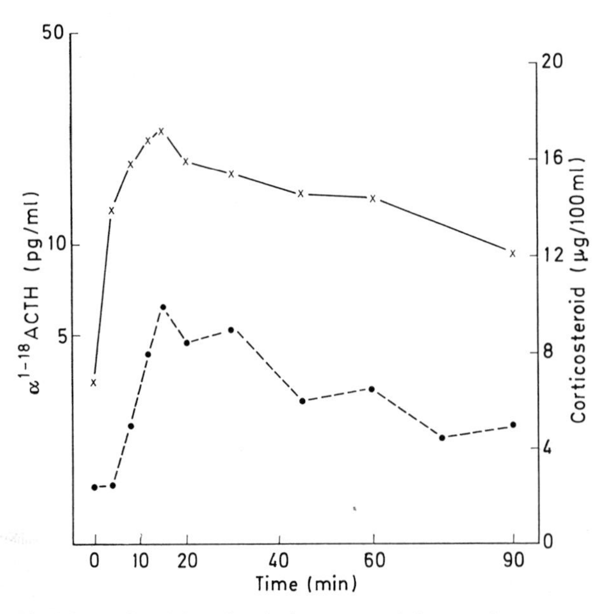

Fig. 3.9. As Fig. 3.8 but after the intravenous injection of only 60 ng of the substituted α_{1-18} ACTH. The same prolongation of the influence of the substituted ACTH occurred even with this low dose, demonstrating that the prolonged half-time was not due to overloading the systems contributing to the removal of the biologically active 'hormone' from the circulation. (Daly et al. 1974a, p.339)

Chapter 4

Maintenance Culture

A. Importance of Organ Culture

I. Changes Induced by Proliferative Culture

It might have been argued that the simplest way of obtaining sufficiently large samples of a particular cell type from one animal was to grow these cells in proliferative culture. This argument is based, inherently, on a number of assumptions which are not supported by biological expectations or experimental facts.

To begin with, anyone who has taken small explants of chick heart and grown them in proliferative culture, will be well aware of the fact that the cultures will consist mainly of fibroblasts. This is one of the standard techniques for growing fibroblasts (e. g. Davies 1954; Willmer 1965 a; Moscona et al. 1965). In more general terms, it is by no means certain that the cells which will grow out in proliferative culture (in this case, fibroblasts) will reflect the predominant cell type of the explant (which in the case cited should have been myocardial muscle cells).

But there is a further consideration. A differentiated cell has properties which characterize that cell type. When it is induced to become mitotically active certain of the genes, which have been de-repressed so that they can exert their influence to establish the differentiated state of the cells, will have to be repressed again. Other genes, namely those required for mitotic activity, will have to be activated (or de-repressed). The cellular activity associated with mitosis is likely to be very different from that essential for the specialized metabolism of the differentiated cells (e. g. Fell 1969). This effect has been shown in relation to chondrocytes (Holtzer and Abbott 1968).

II. Disadvantages of Isolated Cells

Many studies are made on cells isolated from a particular region of a tissue. The region can be separated by dissection, or the cells of the whole organ can be isolated and then the specific cell type can be separated and concentrated by physical techniques. For example, in the studies of Mendelsohn and Mackie (1975, after the method of Haning et al. 1970) the cells of the zona fasciculata and zona reticularis of the adrenal were separated from those of the zona glomerulosa and of the medulla. Indeed there are several valuable bioassays based on such isolated cell preparations (e. g. that of Sayers et al. 1971, for ACTH, using isolated cortical cells from the adrenal; or that of Quazi et al. 1974, in which luteinizing hormone is assayed by its effect on Leydig cells isolated from the testis). One of the most interesting applications of this type of preparation is that in which the cells are bound on to a column so that the effluent from the cells can be measured following hormonal stimulation (e.g. Lowry and McMartin 1974; Hudson and McMartin 1975).

The use of isolated cells does not suffer from the de-differentiation problems discussed earlier, since the cells remain fully differentiated. But, of course, only relatively few cells can be obtained from any one animal. The advantage of proliferative culture was that it increased the number of genetically identical cells (although the genetics too can alter with prolonged culture). The isolated cell procedure must use cells derived from several animals in order to have sufficient cells. Thus although it overcomes the de-repression problems of proliferative culture, it does have the between-animal variability which is likely to diminish sensitivity.

However, the most serious objection to any isolated cell procedure is the fact that the cells have been removed from their natural cellular environment. The treatment required to effect this, whether it is mechanical or enzymatic, can adversely affect the surfaces of the isolated cells. In the extreme example the method of isolating the cells can remove the binding site at the cell surface so that the cells can no longer respond to the activator or do so with diminished sensitivity (Winand et al. 1976). A less extreme example is that of the isolated adrenal cells which cannot respond to ACTH if it is administered in plasma; for the assay of Sayers et al. (1971) the hormone must first be extracted from the plasma.

But the environment of the cell, in the tissue, may be of considerable importance in deciding the sensitivity of the response of the cell. This 'environment' includes not only the cell-to-cell contacts but also the relationship between the cells and their stroma. At present we know too little about how these factors influence cellular response to hormones. There is some evidence of co-operativity between the different cell types in their response to a single hormone. There is some reason for believing that co-operation exists between the cells of the zona fasciculata of the adrenal cortex (which store cholesterol for the ultimate synthesis of other steroid hormones) and of the zona reticularis (which may be the most active site of synthesis of some of these hormones).

Even beyond this type of co-operation between cell types, there is also the possibility that some recognition signal, which may be a wave of membrane potential charge as described by Petersen (1975; also Lowenstein 1966) may also play a major part in the response of a target tissue to a hormone. Such 'signals' are not yet well elucidated. But they would account for the fact that the response of an organized tissue, as in the cytochemical bioassays, can be several orders more sensitive than is the recorded response of the equivalent isolated cells.

B. Earlier Use of Organ Culture

There is a great wealth of information concerning the culture of organs (as reviewed in Willmer 1965 b). Several of the leading figures in this field of endeavour have been very aware of the potential value of such organ culture for studying the effects of hormones. Lasnitski (1965) investigated their effect on explants of the prostate gland maintained in vitro. Gaillard (1955, 1961; Gaillard and Schaberg 1965) was the first to try to quantify the histological responses of embryonic long bones (of mice) to exposure to PTH in vitro. The difficulty is that histological changes are not readily quantifiable and they are likely to be relatively insensitive, requiring either abnormally high concentrations of the hormone or extended periods of time (or both) to become measurable. Consequently in studies on bone resorption, other workers used bones which had been 'labelled' with radioactive calcium before exposure to parathyroid hormone (e. g. Raisz 1963; Reynolds and Dingle 1970) and measured the release of radioactive calcium. However, these methods too do not seem to have been suitable for a statistically valid bioassay technique (Zanelli and Parsons, to be published). Other workers have used similar techniques on other endocrine tissues and apparently reached the same conclusion.

The maintenance organ culture system of Fell and Weiss (1965) has been used in other in vitro assays of PTH, notable by Zanelli et al. (1969; modified by Webster et al. 1974). In this assay, post-natal mouse calvariae are maintained in vitro for a couple of days to achieve equilibration and are then exposed for 2–4 days either to graded concentrations of a standard PTH preparation or to the test material. The amounts of calcium, inorganic phosphate and hydroxyproline liberated into the culture medium show a log-dose relationship for concentrations of the hormone of between 0.01 and 1.0 IU/ml culture fluid (Zanelli and Parsons, to be published). This is the same sensitivity as is given by the radioactive calcium method of Raisz (1963). It is still insufficiently sensitive to assay the circulating levels of the biologically active hormone.

It is reasonable to assume that biochemical changes induced by the hormone would provide greater sensitivity. Indeed many workers have studied the biochemical effects produced when PTH acts on bone in vitro (as reviewed by Zanelli and Parsons, to be published). Yet these studies do not seem to have yielded results which could be used as a basis for bioassay.

C. Maintenance Culture as Used in Cytochemical Bioassays

I. Aims

The primary purpose of maintenance culture as a prelude to the cytochemical bioassays is to have the target tissue of the hormone readily accessible to manipulation and in a state in which it is most responsive to the hormone. The tissue should be divisible into a sufficient number of explants for the response of that particular target tissue to be assessed against graded concentrations of the hormone (so as to provide a dose-response calibration graph for that tissue) and against two concentrations of the sample (e. g. plasma) to be assayed. In this way the assays can be within-animal assays, so obviating the between-animal variations that have contributed to the relative insensitivity of most conventional bioassays.

Removal of the target tissue from the animal, and its maintenance in non-proliferative adult organ culture, also has the following advantages: it frees the tissue from the hormonal influence of the animal and allows the cells to recover from such hormonal stimulation; it allows the cells to recover from the trauma of excision; and, consequent on these factors, it allows the cells to replenish metabolites or restore membrane permeability (or impermeability), and to re-generate enzyme activities to a basal state. These factors may play an appreciable part in the sensitivity of the cytochemical bioassays and they merit fuller consideration; it is possible that the sensitivity of some other in vitro assays would have been enhanced had these factors been recognized.

1. Removal from the Hormonal Influence of the Animal

For simplicity let us take a particular example, such as the cytochemical bioassay of ACTH, which is illustrative of the problems encountered in all the assays. The assay depends on the fact that ACTH depletes ascorbate from the cells of the zona reticularis in a log-dose related fashion.

During life the adrenal cortex is subjected to the effects of this hormone. Therefore at the moment of death the cells will be more or less depleted of ascorbic acid, depending on the circulating concentration of ACTH at that time. This will be true for all hormones. In the case of ACTH there is an additional complication due to the fact that stress (and even humane death is stressful) causes the release of fairly large amounts of ACTH from the pituitary gland. (The effect of stress is used in a test of the hypothalamic-pituitary-adrenal axis, as discussed in Chap. 9). Stress-induced release of ACTH ensures that the adrenal cortex will largely be depleted of its ascorbate content so that the target cells cannot show much more response to ACTH if they are used immediately in the bioassay. Moreover there is some evidence that, with certain hormones, the cells which have been hormonally stimulated may require a short recovery period before they can respond again to the hormone.

Thus, whether stress influences the levels of the hormone under investigation or not, it is necessary to remove the target tissue from the influence of the circulating hormone. This, of course, is well known so that a prerequisite of many in vivo bioassays of pituitary hormones is hypophysectomy. In a sense the removal of the target organ from the animal after death constitutes a form of hypophysectomy which can be achieved by relatively unskilled operators.

The practical consequences of leaving the cells still influenced by their previous exposure to the hormone is seen in the study by Chambers and Chayen (1976). This was an investigation of the effects of ACTH on the 5′-nucleotidase activity of the membranes of cells of the zona reticularis of the adrenal gland. It was shown that if the glands were removed from the guinea-pig, maintained in vitro for 5 h and then exposed to increasing concentrations of ACTH, the enzyme activity rose linearly with the log-dose of hormone from 5×10^{-19} g/ml to 5×10^{-11} g/ml. This showed apparently that the cell membrane responded very sensitively to the hormone. But if, at the end of the maintenance culture, the adrenal tissue had been primed with ACTH at a concentration (5×10^{-16} g/ml) which was too low to be discernible in the normal ascorbate-depletion assay, the 5′-nucleotidase response following further exposure to the hormone was biphasic. There was a linear positive correlation between enzyme activity and hormone concentration (after priming) at concentrations of 5×10^{-19} to 5×10^{-17} g/ml but with concentrations greater than the priming concentration the response was linear but negative. Thus at these higher concentrations (5×10^{-17} to 5×10^{-12} g/ml) increasing concentrations of ACTH evoked decreasing activities of the enzyme. The priming dose was given 4 min before the end of the maintenance culture, so that its effect would have been maximal when the tissue was chilled, i. e. just prior to the second exposure to ACTH. And we know that, normally, the maximal response to ACTH is followed by a recovery period (as shown in Chap. 15, Fig. 15.1). Consequently it seems likely that the peculiar biphasic effect was due to interaction of the hormone with the recovery phase of the cell membranes. A similar phenomenon is likely to occur when the normal circulating hormone, in life, takes the place of the priming dose in these studies. And indeed such effects have been found when target tissues are removed from the animal at death and used directly for section assays. For example, it was thought that the gastric fundus could be taken directly from fasting guinea-pigs and used without prior maintenance culture for a section assay of gastrin, seeing that their circulating levels of gastrin would be very low; however, anomalous results of the type described in connection with the effect of a priming dose of ACTH were found.

The use of maintenance culture to remove the target tissue from the hormonal influence of the animal has an additional advantage when a new cytochemical bioassay is to be developed. For example, when studies were begun on PTH, it was found that kidney explants could be maintained in vitro and that they showed reasonably good histological preservation. But it was noted that the activity of glucose-6-phosphate dehydrogenase in certain regions of the tubules was markedly depressed. Two interpretations of this observation were possible: either that the culture conditions, which may have been sufficient for the moment to keep the histological structure, were inadequate for the normal metabolic activity of these cells; or that this enzyme activity was under direct hormonal influence and therefore decreased in the absence of the hormone. The first interpretation was made less likely by the fact that several other enzymes were as active in the cultured tissue as in the tissue which was removed and chilled at the death of the animal. Treatment of the tissue in vitro with PTH showed that this hormone was capable, even at low concentrations (e.g. 0.1×10^{-12} g/ml) of restoring this enzymatic activity; this observation opened the way to a possible bioassay of this hormone (Chambers et al. 1976).

2. Recovery from the Trauma of Excision

The acts of removing a tissue from an animal and cutting it up into explants of suitable size inevitably induce some degree of trauma in the cells. This was shown very clearly many years ago in some studies made in collaboration with the originator of the culture method used in these cytochemical bioassays, the late Dr. O.A. Trowell. We took small explants from rat liver, cut with a guillotine to minimize the trauma of cutting, and maintained them in vitro. Over the first few hours of maintenance culture the cells, studied in cryostat sections by conventional histological procedures, showed remarkable intracellular vacuolation and they lost their glycogen content even though the culture medium also contained insulin. Recovery began after about 3 h in culture and by 5 h the cells looked as normal as did those in the biopsy sample; they had also apparently regenerated their glycogen content. In other studies it has been found that the adrenal cortex is unresponsive to ACTH if it has been in maintenance culture for only 3 h but that full sensitivity returns by the end of 5 h in culture (N. Loveridge, unpublished results).

3. Restoration of Metabolities

The period of maintenance culture is normally 5 h. Obviously if the effect of the endogenous hormonal stimulation persists for longer than this period, an extended time in culture will have to be used. But generally this time is sufficient. Thus the effect of TSH on the lysosomal membranes of the thyroid follicle cells seems to have worn off by this time, in that the membranes show considerably more impermeability after this period of culture than they do in the biopsy sample, when the tissue was under TSH influence (as discussed in Chap. 10). But some systems require that the culture medium be reinforced with essential metabolites. This was seen with the adrenal gland explants initially by the fall in activity of the two dehydrogenases of the hexose monophosphate pathway. As has been discussed above with respect to a similar effect induced by the lack of PTH in the kidney explants, this loss of activity (Table 3.2) could have been due to the inadequacy of the medium or to lack of a regulating hormone. Seeing that it is known that the adrenal gland contains relatively large amounts of ascorbate, it seemed reasonable to assume that this loss of enzymatic activity was due to an inadequate supply of ascorbate. And indeed this appeared to be the case because the addition of ascorbate at a concentration of 10^{-3} M restored full enzymatic activity (Table 3.2). Of course this was fortuitous because the ascorbate levels in the adrenal cortex of different animals varied considerably (Table 3.3), presumably associated with the degree of stress at death. (But it is noteworthy that such fluctuations were much smaller in the cells of the zona fasciculata, as shown by Chayen et al. 1972a.) Incorporation of a fairly high concentration of ascorbate in the culture medium ensured a high concentration in the cells when they came to be exposed to the hormone, so adding to the sensitivity of the ascorbate-depletion assay. However, in relation to the time needed to recover from the trauma of excision (as discussed above), it is of interest that the full replenishment of ascorbate does not occur after 3 h in culture, but only after 5 h (Table 4.1).

In other systems it may be found that other hormones are essential for preparing the cells to disclose their full response to the selected hormone. Thus it may be that the full sensitivity of the response of gastric parietal cells to gastrin requires that the cells be first exposed to acetyl choline. Or it is conceivable that the cells of the corpus luteum may need to be primed by oestrogen before their full response to luteinizing hormone can be elicited.

So it can be seen that the task of elucidating the conditions of maintenance culture which are ideal for a particular tissue can be very rewarding in gaining insight into the endocrinology, physiology and biochemistry of that tissue. When sufficient understanding has been achieved, and the cells are maintained in a relatively ideal physiological state, then a study of the effects of the selected hormone can lead to an adequate bioassay.

Table 4.1. Ascorbate levels in the zona reticularis ($E_{680}-E_{480}$/unit field)[a]

Specimen	Time (h) in culture					
	In T8 medium	In T8 + 10^{-3} M ascorbate				
	5	0	1	3	5	6
1	10.8	–	–	–	22.2	–
2	20.3	–	–	–	23.8	–
3	17.5	19.7	18.8	16.1	22.0	22.5

[a] 1 and 2, recalculated for extinction, Chayen et al. (1972a). 3, Chayen et al. (1976)

II. Procedure

The details of how samples of target tissues are explanted into adult organ maintenance culture of the type developed by Trowell (1959) are to be found in Appendix 1. Here we will be concerned rather with the general outline of the procedure, as used in the cytochemical segment assays, namely in those assays in which the hormone is made to act on whole segments of the target organ.

1. Selection of the Animal Species

For most of the assays developed up to now the animal has been the guinea-pig. There has been no obvious advantage from either sex. The age and body weight may be significant: the thyroids of older guinea-pigs frequently are more cystic and consequently younger guinea-pigs of body weight about 250–300 g are preferred. On the other hand, the adrenal glands of older animals are larger so that guinea-pigs of about 400–500 g are selected for the ACTH bioassay.

2. Segments

The animals are killed by asphyxiation in nitrogen. The target organ, e.g. the thyroid gland for the TSH assay or both adrenal glands for the assay of ACTH, is removed as quickly and as dextrously as possible. The organ is freed of any adherent fat or connective tissue.

Each adrenal gland is divided into three segments, making six in all. It has been found that each gland gives similar values for the biochemical activities which have been investigated and for their content of ascorbate; they show similar responses to ACTH. Consequently all six segments can be used as if they were equal parts of one organ. The thyroid gland of the guinea-pig can also be treated as a single organ and is cut into six segments, three from each lobe. The parietal cells of the stomach occur in the fundus, which is cut free from the rest of the organ, cleared of debris and cut into a series of strips. Thus from each organ at least six segments are cut. Each is maintained separately in vitro.

3. Culture Medium

The basic culture medium is Trowell's (1959) T8 medium. This is composed of all the essential amino acids (except for glutamine, which has to be added just prior to use because this amino acid readily breaks down in solution) and a balanced mixture of salts giving a final tonicity of 325 mosmol per kg water, which is close to that of blood. Although it originally contained antibiotics, these are not usually included in commercially available culture fluids and are not used in these assays because of their potentially inhibitory effects on certain cellular biochemical activities, notably the synthesis of protein and of RNA (e.g. Garren et al. 1971; Schulster 1974). The medium also normally contains insulin, and this has been retained. Bicarbonate is present in the T8 medium at such a concentration that it will achieve a pH of 7.6 when equilibrated against 5% carbon dioxide (with 95% oxygen or air). The medium also contains neutral red as an indicator of pH.

As has been discussed previously, different tissues may require different additions to the medium. To maintain adrenal gland segments, ascorbate (10^{-3} M) is added to the medium. It may be required to maintain stomach segments at a pH of 7.0; to achieve this, a calculated amount of acid is added to the medium so that the new concentration of bicarbonate, equilibrated against the same mixture of gases, buffers at this pH.

4. Maintenance Culture

Each segment is placed on a defatted lens tissue lying on an open mesh metal table (Fig. 4.1). The table stands in a vitreosil glass dish and culture medium is added up to the level of the lens tissue. The vitreosil glass dish is then placed in a Perspex outer chamber which is sealed by a layer of lanoline between the flat top and the chamber walls. The whole assembly is then transferred to a constant temperature room maintained at 37° C or slightly lower. The air in the chamber is displaced by the requisite gas mixture (i. e. 95% oxygen: 5% carbon dioxide) by passing the mixture through a wash bottle in which the water is at 37° C (or thereabouts), and thence into the chamber, for about 2 min. The entry and exit tubes are then closed off by the use of tapered conical spigots which insert readily into the tubes. The tissues are then left for 5 h at 37° C (or slightly below this temperature). The colour of the indicator in the medium should remain unchanged, indicating that there was no leakage of gases.

At the end of this period the segments should be ready for exposure to the hormone, as will be described in turn for each hormone (in later chapters). The whole assembly is shown in Fig. 4.2.

III. Technical Considerations

It will be apparent that the segment of tissue lies above the medium, fed by the lens tissue. Thus it is freely exposed to the gaseous phase and is not immersed in the culture medium; immersion into a fluid has been found to be deleterious to the full survival of tissues.

The gaseous phase used in all the assays up to the present has been 95% oxygen: 5% carbon dioxide. It has often been said that a gaseous phase containing 95% oxygen can be toxic to cells. On a number of occasions, with diverse tissues, my colleagues have compared the effects of 95% air: 5% carbon dioxide, or of air: nitrogen: carbon dioxide, with those of our standard mixture containing 95% oxygen: 5% carbon dioxide, and in all cases the last gave better preservation whether judged histologically or cytochemically. This may not apply to all tissues; for example 5% carbon dioxide in air, not oxygen, is said to be required by certain types of bone cultures (as reviewed by Zanelli and Parsons, to be published).

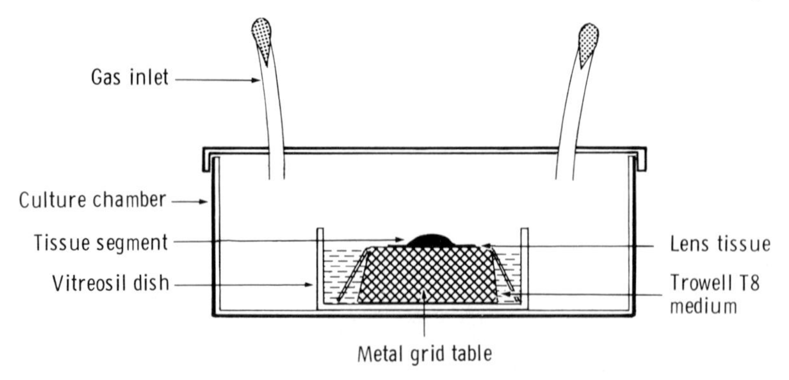

Fig. 4.1. Diagram of a culture pot used for maintaining tissue in vitro

Fig. 4.2. The culture system. The culture pot (on the *left*) is connected to the gas line, with the 95% O_2 : 5% CO_2 passing from the gas cylinder through a wash bottle before entering the culture pot. The whole assembly is in a room maintained at 37° C

Chapter 5

Cytochemical Methods

A. Basic Procedures

The theoretical basis of cytochemical procedures, and details of the practical methods, have been fully reviewed in a number of recent publications (e. g. Chayen et al. 1973 b, 1975; Bitensky and Chayen 1977; Chayen 1978 a, b). However, those aspects which are particularly relevant to the cytochemical bioassay of hormones will be reviewed in this chapter to make it easier to understand the chemical basis of the assay systems.

As discussed in Chap. 1, the aim of cytochemistry, or cellular biochemistry, is to extend rigorous biochemical analysis to the cellular level and so allow biochemical activity to be related to histology. To achieve this end, certain modifications had to be made to the procedures used in conventional biochemistry (as indicated in Table 1.1).

I. Preparation of Tissue Slices

1. Histological Techniques

One of the aims of cytochemistry is to relate biochemical activity to histology. This requires that, in the place of the biochemist's tissue slice (as discussed in Chap. 1), we use thin tissue slices which are sufficiently thin for the histology to be defined, e. g. by phase-contrast microscopy. The upper limit of thickness imposed by this requirement will depend on the dimensions of the cells in the tissue. Thus in liver, in which the hepatocytes may be 30 × 25 × 25 μm, the thickest section which will still allow a clear view of the histology, with relatively little overlap of cells one above the other, may be about 25 μm; in other tissues, with cells of more normal size, the section should not be thicker than 10–18 μm.

The basic problem in cutting sections of tissue is that the block of tissue must be rendered firm so that it does not 'give' or flow around the knife. In conventional histology, this is achieved by impregnating the tissue with a hard matrix, the most commonly used being a form of paraffin wax. In electron microscopy, where the sections must be much thinner (e. g. 100–50 nm thick), the matrix must be considerably harder and a methacrylate or Araldite resin is used. A double-embedding procedure, utilizing first a paraffin wax and then a methacrylate resin, is also used in conventional histology when relatively thin sections (e. g. 1 μm thick) are to be cut of a delicate tissue.

Paraffin wax is solid at normal laboratory temperatures so it has to be heated and infiltrated into the tissues in the hot, fluid state. And the hydrophobic wax is not miscible with water. Thus, if fresh tissue is immersed in hot molten wax, the result is disastrous. This is overcome, in conventional histological technique, by dehydrating the tissue, usually by the use of alcohol, then infiltrating it with a hydrophobic solvent such as chloroform or xylene with which molten wax is readily miscible. To protect the tissue against deleterious effects of dehydration, of the solvent effect of chloroform or xylene, and of the solvent and mechanical influences of the molten wax, the tissue is first fixed in a chemical, or mixture of chemicals. This chemical fixative may be a precipitant or it may more gently coagulate the protoplasm. The physical effects of different chemical fixatives have been discussed fully by Baker (1945, 1958). Many fixatives cause cross-linkage between proteins, so inactivating the active groups of such proteins (e. g. Wolman 1955); all are designed to denature the protoplasm.

Thus the conventional histological procedures involve the following steps:

1) Chemical fixation which denatures protoplasm, inactivates active groups and so tends to destroy enzymatic action. It 'fixes' the protoplasm against the ravages of the later procedures and stops autolysis – which means that it inactivates enzymes which could cause the degradation of the cells. Some histological fixatives are specially preferred because they also 'mordant' the protoplasm so that the sections stain brilliantly with histological dyes. This mordanting effect is of two types: a true mordanting of the protoplasm which involves the binding to it of heavy metals which then can bind histological dyes, or merely the unmasking of charged groups, in proteins or nucleic acids or carbohydrate moieties, which can then be coloured by the electrostatic binding of a dye.

   All these effects, admirable as they are for producing a well-stained histological section, are inimical to the study of the cellular biochemistry of the cells in that section. The inhibition of enzyme activity caused by various chemical fixatives is listed in Appendix 1 of Chayen et al. (1973 b, 1975); the changes in the active groups of structural components, such as the freeing of otherwise bound phosphate moieties of nucleic acids, or the effects on phospholipids in membranes, are incalculable.
2) Dehydration, usually through graded concentrations of ethanol up to absolute ethanol.
3) Infiltration of xylene, chloroform or some similar solvent which first mixes with the absolute alcohol in the tissues. It is well known that mixtures of methanol and chloroform are the best solvents for lipids because they first break phospholipid-protein complexes, whether they occur as lipoproteins or as proteolipids, and then extract the now-free lipid components; it is likely that this step of the histological procedure would be nearly as effective in disturbing lipid-protein associations.
4) Infiltration whith molten paraffin wax, which itself is an adequate solvent for some lipids.

It has been shown that there can be considerable loss of enzymatic activity when fixed tissue, which already has lost appreciable enzymatic activity, is subjected to steps (3) and (4) (Appendix 1 of Chayen et al. 1973 b, 1975). The loss of lipids and of nitrogenous matter has been less fully evaluated but seems to be considerable (Ostrowski et al. 1962 a, b, c). Although various modifications to this general scheme have been tried, such as simultaneous fixation and dehydration in cold acetone, or freeze-substitution, in which the dehydration takes place at lower temperatures, the gains have not been sufficient to merit consideration as procedures for cellular biochemistry (see Chayen 1978 a). Many studies were also made on the possibility of freeze-drying tissues and then infiltrating molten paraffin wax into the dry protoplasm (e. g. Bell 1956). There seem to have been problems in drying the tissue without causing ice artefacts, and it has been shown that molten wax, acting on frozen-dried tissue, causes marked loss of enzyme activity (as reviewed in Appendix 1 of Chayen et al. 1973 b, 1975).

2. Chilling

The other way in which a sample of tissue can be rendered sufficiently firm to allow it to be sectioned is by cooling it. In the simplest procedures the tissue is frozen and sectioned in the frozen state. This obviates the loss of activity due to chemical fixation but it has its own disadvantages, namely the structural and chemical damage caused by the formation of ice within the tissue. This subject has been much investigated (e. g. Bell 1956; Luyet 1951, 1960; Asahina 1956; Chayen and Bitensky 1968).

When tissue is cooled slowly the extracellular water is the first to freeze and forms ice crystals in the extracellular spaces. As a result of the increasing tonicity of the aqueous solution remaining outside the cells, intracellular water is drawn out of the cells by exosmosis. The cells shrink and the tonicity of the cytosol increases. Increased tonicity causes rupture of lipid-protein complexes and so destroys the integrity of membranes (Lovelock 1957), both those bounding the cell and those of the subcellular organelles. Finally, with prolonged cooling at a sufficiently low temperature, all the water will crystallize as ice, causing the cell structure to become mechanically shattered.

Yet it has been known for some time (e. g. Asahina 1956; Luyet 1951, 1960) that protoplasm can be cooled to very low temperatures without allowing the water to crystallize out of the colloidal sol state to form ice. This phenomenon is also well known in other branches of science, particularly in the study of clouds (Knight 1967) where water itself may occur in a super-cooled state. In brief, water is a highly complex system of molecules, some of which are held together and some of which are relatively free. The water molecules or molecular species are in an amorphous state in fluid water (or in steam). When they are cooled rapidly they lose mobility and if nothing untoward is present, such as centres of crystal nucleation, they can merely lose their mobility with increasingly lowered temperature until they become immobile, set in their amorphous distribution. Crystallization, into ice, requires energy (usually, but not necessarily, thermal) to overcome the electric repulsion between the charged micelles of water molecules. It is helped if there are centres which can form the nucleus for the growth of crystals. Thus it is possible to have super-cooled water, at temperatures well below 0° C, even in a mass of water such as a cloud. In protoplasm there is little 'pure' water since most tissue water has some colloidal matter, such as protein, dissolved in it, and such colloids tend to diminish the effects of potential centres of nucleation. Moreover, as was shown particularly by Bungenberg de Jong (1936), much of the water of protoplasm occurs in coacervates in which the water molecules are bound in a complex colloidal structure (Fig. 5.1). These structures retard the crystallization of the bound water into ice.

Thus in the studies of Lynch et al. (1966) it was not surprising to find that if a thermocouple was inserted into a block of tissue and the tissue was cooled under

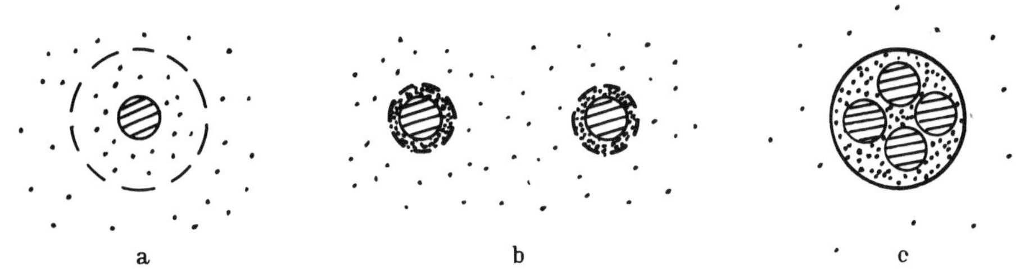

Fig. 5.1 a–c. Diagrammatic representation (after Bungenberg de Jong 1936) of states of water in relation to various particles (shown as *hatched circles*). **a** The water molecules tend to aggregate around the particle, to from a hydration layer (*broken lines*), but leaving the particle suspended in an aqueous medium; **b** Two such particles, with dense hydration layers, still suspended in an aqueous environment; **c** Such particles can come together; the hydration layers coalesce to form a coacervate of several particles within a common hydration boundary

suitable conditions to −40° C or below, there was no sign of the formation of ice in the tissue (Fig. 5.2). Indeed, in these studies, ice was observed when such super-cooled tissue was allowed to warm-up to about −5° C at which the thermal energy was sufficient to allow crystallization. Thus it is possible to cool water to, say −105°C, without precipitating ice and then to form ice by heating that tissue by 100° C (to −5° C).

The various temperatures which most favour the formation of ice from water, or of one form of ice from another, were defined by Tammann (1900). From this information (Table 5.1) is seems clear that if tissue can be cooled to below −40° C without allowing the water to crystallize, then there is a reasonable prospect of keeping it in the amorphous, chilled (super-cooled) state. From his figures it is noteworthy that temperatures of about −18° C are dangerous; it is unfortunate that many who use freeze-sectioning methods tend to favour such temperatures.

3. Chilling Procedure Used in Cytochemical Bioassays

Small pieces of tissue, which may be as large as 5 × 5 × 3 mm, are chilled by precipitate immersion into *n*-hexane at −70° C. The hexane must be free from aromatic hydrocarbons; its boiling range should be 67°–70° C which is about 140° C above the temperature to which the tissue is chilled.

It might have been thought better to chill the tissue in liquid nitrogen, which is at −193° C. But this temperature is at, or very close to, the boiling point of nitrogen. Consequently this procedure, in fact, causes the tissue to cool only relatively slowly

Table 5.1. Triple points of ice mixtures (Tammann, quoted by Chayen and Bitensky, 1968)

Phase	Temperature of the triple point (° C)
Ice 1 – liquid – vapour	+0.0075
Ice 5 – liquid – ice 6	+0.16
Ice 3 – liquid – ice 5	−17
Ice 1 – liquid – ice 3	−22
Ice 2 – ice 3 – ice 5	−24.3
Ice 1 – ice 2 – ice 3	−34.7

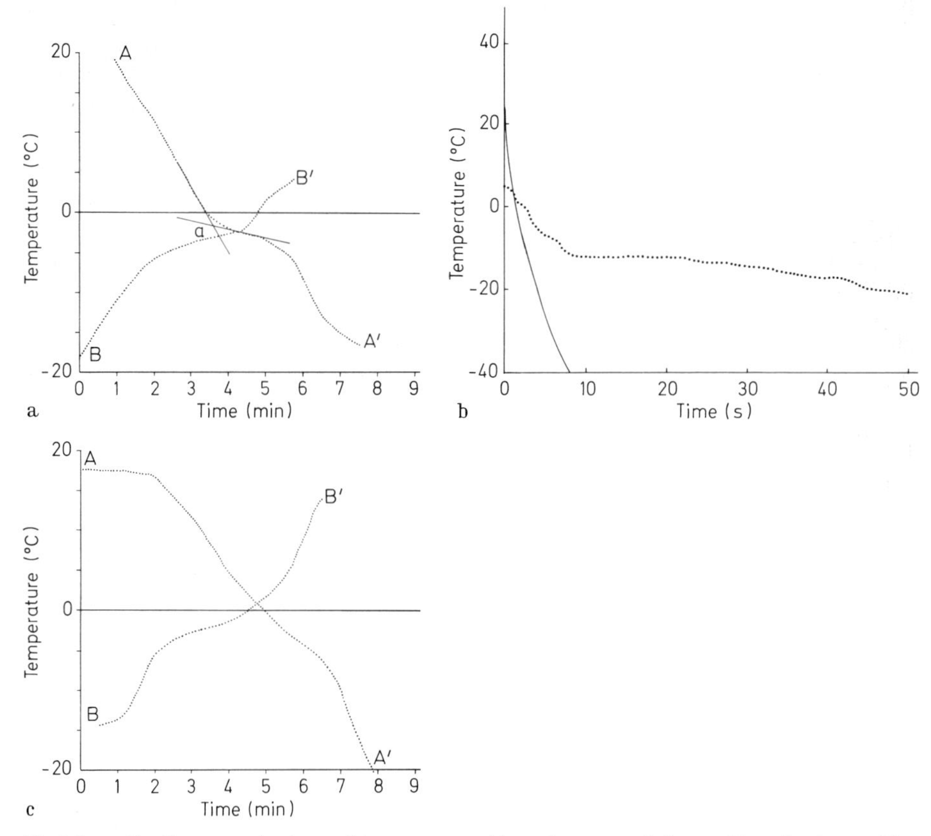

Fig. 5.2 a–c. Cooling curves in pieces of tissue, measured by a thermocouple inserted into the tissue. **a** The formation of ice: a large piece of rat liver was cooled (A to A′) and then warmed again (B to B′). During cooling (A to A′) there was a marked deviation from linearity at temperatures of between 0 and −5° C, denoting the formation of ice. The intercepts at **a** suggest that ice began to form at around −1.3° C. The curve obtained when the same tissue was warmed (B to B′) was almost symmetrical with the cooling curve, suggesting that the same amount of ice, in the tissue, was being thawed. (Lynch et al. 1965, p. 217). **b** A small piece of rat liver chilled in hexane (*solid line*). There is no deformation of the cooling curve, indicating that no detectable ice formed during the cooling. The broken line shows the cooling curve of a large piece of muscle, with obvious change in the cooling rate due to the formation of ice. (Chayen et al. 1973 b, p. 4). **c** The cooling curve (A to A′) of this piece of rat liver showed only slight deviation from linearity during cooling. This might correspond to the formation of a small quantity of ice or, more likely, it may be instrumental. In contrast, the warming curve (B to B′) showed a considerable change in shape, beginning below −10° C, apparently due to the formation of a considerable quantity of ice caused by warming the tissue. (Lynch et al. 1965, p. 216)

because when the warm tissue enters the liquid nitrogen its heat causes the nitrogen to boil around the tissue and so insulate it with a blanket of gaseous nitrogen (e.g. see Moline and Glenner 1964; Chayen et al. 1973 b). Thus it is necessary for the coolant to be at a temperature far removed from its boiling point. It should also be well above its freezing point because the conduction of heat depends on the contact between the coolant and the specimen and this is best achieved by a fluid. The coolant should also be a good conductor of heat and have a high thermal capacity.

The hexane, in a beaker, is kept at −70° C by an outer bath containing crushed solid carbon dioxide in absolute ethanol (or industrial spirit). The tissue is left in it for up to 1 min and is then removed by the use of precooled forceps (to ensure that the heat in the forceps will not thaw the tissue); excess hexane is quickly shaken off from the specimen which is deposited in a dry glass tube (3 inches by 1 inch size). The tube has been cooled to −70° C by standing in a Dewar (vacuum) flask in intimate contact with solid carbon dioxide. It may contain a layer of absorbent paper in the bottom of the tube to remove surplus hexane; if so, this paper will have been in the tube in the Dewar flask for some time previously to ensure that it too is sufficiently cold. When the tissue specimen has been dropped into the dry tube, the latter is corked and replaced quickly into the packed carbon dioxide ice. The handling of the tissue must be done expeditiously to protect it from thawing.

Even when the tissue is stored at −70° C it will deteriorate. Hence for the segment assays it should not be kept for longer than 1 week under these conditions; for the section assays it should be used within 3 days of chilling.

4. Mounting the Specimen

The chilled tissue must be mounted on a microtome chuck before it can be sectioned. This is a potentially hazardous procedure because of the danger of thawing. In practice a cavity is prepared in a block of solid carbon dioxide into which the specimen can be inserted and covered with another piece of carbon dioxide ice. A metal microtome chuck with a flat (usually grooved) top is stood in a slurry of crushed carbon dioxide ice in absolute ethanol (or industrial spirit) at −70° C. When it is cold, water is added to the top of the chuck to form a frozen platform. The tissue specimen is then inserted into the cavity in the block of solid carbon dioxide and its orientation is defined. It is then covered by another piece of solid carbon dioxide. Another drop of water (often coloured with toluidine blue to act as a background for the specimen) is added to the ice platform and, when there is only a thin surface film of water left the specimen is removed from the cavity in the block of solid carbon dioxide and placed on that film. The film of water freezes, holding the tissue on to the ice platform. At all stages, when the tissue is to be handled, the forceps must be precooled to avoid warming the specimen.

The portion of the tissue close to the water film obviously cannot be used because the warmth of the water will have heated the tissue at this point. However, if the procedure is done quickly, the rest of the specimen above this region will be free from ice. The hazards of the whole technique can be minimized if it is done inside a cryostat cabinet at −25° C or −30° C. They can also be reduced if a concentrated solution, such as a homogenate of rat liver, is used instead of water. The chuck is removed from the coolant slurry, wiped dry and left in the cryostat cabinet until the tissue is to be sectioned.

5. Sectioning

The tissue has been hardened by chilling and therefore is receptive to sectioning. This is done by means of a cryostat microtome (Fig. 5.3 a). This consists of a conventional microtome set in a cryostat, namely a box which can be maintained at a low temperature.

The act of cutting a slice off a chilled block is liable to generate heat both from the energy of impact and from the friction of the knife through the block. This is

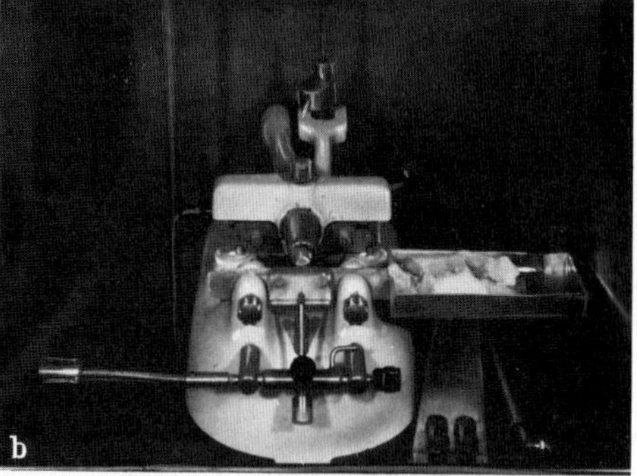

Fig. 5.3 a, b. The cryostat microtome. **a** This instrument consists of a conventional microtome set in a refrigerated cabinet (here shown with the front window open, and turned back over the top of the cabinet). Sections can be cut either manually, by turning the handle on the right of the photograph, or by the automatic cutting device, on which the handle is mounted. **b** This view of the microtome, inside the refrigerated cabinet, shows the cradle used for packing solid CO_2 around the haft of the microtome knife. The guide plate, used to keep the sections flat, is resting on the knife

overcome (1) by cutting at a low ambient temperature: $-25°$ to $-30°$ C is recommended; and (2) by having the knife cooled to below this temperature by packing solid carbon dioxide around its haft (Fig. 5.3 b). Any heat generated by the sectioning procedure will be dissipated into the knife which is a better conductor of heat than is the tissue, and is at a considerably lower temperature (just below $-70°$ C). Silcox et al. (1965) showed that under these conditions no ice crystals and no ice damage could be detected by normal histological examination. Furthermore Altman and Barrnett (1975) cut such sections and then embedded them in methacrylate; examination of ultra-thin sections by electron microscopy showed no signs of ice or of artefact attributable to freezing. But Silcox et al. (1965) also showed that after cutting sections which were free from ice, they could produce ice artefacts in serial sections of the same tissue if they increased the temperature of the cryostat cabinet to $-15°$ C (as preferred by many histologists) or did not cool the knife.

For cytochemical analyses the thickness of the sections must be constant. Yet it is well known in histology that the thickness of sections can vary quite widely. This variability was confirmed by Butcher (1971 a), who used a chemical procedure for determining the thickness of cryostat sections. But he also demonstrated that the thickness varied according to the speed of cutting. When the advance mechanism of the microtome was set to give a nominal thickness of the sections of 14 μm, sections of a block cut slowly were 19.3 μm thick; those cut at a 'normal' speed were 14.2 μm thick while those cut rapidly were 9.7 μm thick. For this reason it is recommended that the microtome be fitted with an automatic cutting device which cuts at a constant speed (Fig. 5.3 a). It is this constancy of thickness which, in turn, yields a constant amount of reaction product per unit area of the section (measured by microdensitometry) which accounts for much of the precision now achieved by cytochemistry.

6. Flash Drying

Up to this stage in the procedure used for preparing sections, the technical steps have been fairly predictable. The real difficulty lay in how to remove the section from the knife. If the theoretical arguments were correct, then we have a section of super-cooled protoplasm lying on a metal knife at about −70° C. Suppose we were to transfer the section with a cold brush on to a cold slide (or cold cover-slip). There would be wrinkles and possibly flaws in the section, but that might be a safe manoeuvre. But the moment that the section was taken out of the cryostat cabinet, the warmth of the laboratory would provide sufficient thermal energy for the super-cooled water to set to crystalline ice (very much as in Fig. 5.2c). The way round the predicament is to flash-dry the section. This is achieved by bringing a glass slide, from the ambient temperature of the laboratory (for some purposes the slide may be warmed even to +40° C) close to, and parallel to, the section on the knife. When the slide is about 2 mm from the section, there is a gradient of over 90° C across the section (from below −70° C at the surface of the knife to +20° C at the surface of the slide). This is sufficient to make the water in the section boil off the section on to the knife and this phenomenon jet-propels the section on to the slide where it adheres because of the speed of impact. The water can be seen to freeze on the knife, leaving a complete imprint of the section. The relatively dry section can safely be withdrawn from the cryostat cabinet into the laboratory because there is now insufficient free water to form ice.

The practical details of this form of cryostat microtomy, including the use of the anti-roll guide plate which keeps the sections flat on the knife while they are being cut, are given in full by Chayen et al. (1973b, 1975). The proofs that these procedures do not cause measurable artefact have been discussed by Chayen and Bitensky (1968) and by Chayen (1978a). The most cogent argument is the response of sections in the cytochemical section bioassays, which will be considered later.

7. Protection of Undenatured Sections

When sections, 10–20 μm thick, of undenatured protoplasm are immersed in an aqueous medium at a relatively neutral pH (between pH 7.0 and 8.5) the undenatured protein and nucleoprotein will dissolve, or become dispersed, into the medium. This is a corollary of the fact that the protoplasm has not suffered denaturation. And indeed various workers showed, by microchemical analysis of tissue sections before and after immersion in such media, that up to 70% of the total nitrogenous content of fresh sections was lost into such aqueous solutions as are used for the biochemical or cytochemical analysis of enzymatic reactions (e.g. Altman and Chayen 1965). This loss of material is not found when the sections are immersed in an acetate buffer at an acidic pH, such as pH 5, or in a very alkaline medium as is used for investigating alkaline phosphatase activity (e.g. pH 9.0). But, since the activity of most oxidative enzymes is maximal between pH values of 7.0 and 8.5, the loss of much of the tissue section during the reaction would totally invalidate the claim that cytochemistry relates biochemical activity to histology: most of the biochemical activity of those oxidative enzymes that are not tightly bound to cellular structures will be in the reaction medium, not in the cells. Any deposition of reaction product in cells either will be fortuitous or will reflect a secondary reaction. The latter occurs when the dehydrogenase enzyme is 'solubilized', reacts with the substrate and coenzyme in the reaction medium and the reduced coenzyme is then reoxidized by a tissue-bound 'diaphorase'. In such a case the reaction product indicates the site and activity of the diaphorase, which may be the microsomal respiratory pathway or the NADH-oxidizing system of mitochondria, and which may be active in cells which do not contain the particular dehydrogenase enzyme which is being studied. (For a review of this problem, which has greatly exercised histochemists, see Chayen 1978 a).

This obstacle to the use of undenatured sections is overcome by the inclusion of a suitable concentration of a colloid stabilizer in the reaction medium. The first to be used was a particular grade of polyvinyl alcohol: when M05/140 Polyviol was included in the reaction medium at a concentration of 20% (w/v) it completely retained all the nitrogenous matter of the sections (Altman and Chayen 1965). (More recent varieties of polyvinyl alcohol may require to be added at slightly greater, or less, concentrations to have the same effect). Moreover it was found that so-called soluble dehydrogenases, such as glucose-6-phosphate dehydrogenase, which biochemists find in the supernatant fractions of their homogenates after differential centrifugation, could also be retained at full activity even when the medium was replaced by fresh medium (Altman and Chayen 1966).

More latterly other colloid stabilizers have been found to have similar properties. Butcher (1971 b) found that a polypeptide derived from the controlled partial degradation of collagen, available commercially as Polypeptide 5115 (Sigma), was equally suitable. Ficoll has also been used for this purpose (Stuart and Simpson 1970) and more recently it has been found that very low concentrations of gum tragacanth may be equally effective. The use of these stabilizers has made it possible to measure cytochemically the activity of enzymes which have pH optima in the range of 7.0–8.5. But because they can stabilize the content of undenatured sections immersed in aqueous media at these pH values, at which many polypeptide hormones are most effective, they have a particular function in the cytochemical section assays.

The way these colloid stabilizers produce their effect is none too certain. Polyvinyl alcohol was selected initially for a number of properties: (1) it is much used to stabilize colloids, such as lipstick, and seems to have no deleterious effect on the biological tissues to which it has been applied; the same pertains, of course, to various gums which are used to stabilize such comestibles as Turkish delight. (2) It had been used to slow down or stop small organisms so that they could be studied by normal or phase-contrast microscopy, but the organisms regained their full motility when the polyvinyl alcohol was removed. (3) It had proved an effective stabilizer of isolated cells (Chayen and Miles 1953), retaining within them soluble components of small molecular weight (Chayen and Jackson 1957). (4) It stabilized subcellular fractions during and after homogenization (e.g. Chayen and Denby 1960).

Two explanations, which are not mutually exclusive, of how such stabilizers produce their effects are (a) by asserting a double-electric layer around the colloidal matter or particles (Sherman 1955) and (b) by the excluded volume mechanism (Scott 1974). The latter accounts for the influence of many long-chain hydrophilic molecules, such as proteoglycans in synovial fluids and in cartilage, in binding water so that it is less available for the solution and flow of small molecules.

B. Demonstration of Dehydrogenase Activity

The general concepts involved in the use of tetrazolium salts for measuring dehydrogenase activity have been discussed in Chap. 1 (Sect. B. III.d). Here we will be concerned only with specific points and methods which are relevant to the present cytochemical bioassay systems.

I. Nature of Reduced Tetrazolium Salt

The nature of the various mono-tetrazolium and di-tetrazolium salts has been discussed by Altman (1976). Neotetrazolium chloride, which is used in the cytochemical bioassays, is a di-tetrazolium salt (Fig. 1.3) so that it can be half-reduced or fully reduced. In the former state, when one molecule of neotetrazolium accepts two atoms of hydrogen, it gives rise to a red formazan; in the latter, it produces a purple di-formazan, due to the reduction of the molecule by four atoms of hydrogen (Altman and Butcher 1973; also Altman 1974). In cytochemical reactions either form may be produced; frequently the earlier reaction product is the half-reduced molecule. However, as was shown by Butcher (1972) and by Butcher and Altman (1973), the absorption curves of the two formazans (the mono- and the di-formazan) as they occur in sections cross at 585 nm. If this isobestic point is used for measuring the dehydrogenase activity then it is possible to calculate the amount of dehydrogenation irrespective of the nature of the formazan which is produced. The molar extinction coefficients of these formazans when present in sections have been determined (Butcher and Altman 1973) so that it is possible to express the microdensitometric measurements of the formazan deposits in absolute terms of moles of hydrogen produced by unit mass of the specific cells in unit time. As a check on such measurements, the contribution due to the red half-formazan can be measured (and calculated) separately by measuring at 520 nm, and that of the purple diformazan at 620 nm (as described by Butcher 1972). This use of a two-wavelength procedure to analyse the amounts of each chromophore in a mixture has been discussed by Chayen and Denby (1968).

II. Terminology

It is probably rigorously correct to speak of a dehydrogenase as removing 'reducing equivalents' from the oxidized substrate and passing these on to the coenzyme or its equivalent. It has become fashionable to talk of the transfer of electrons when discussing oxidation-reduction steps such as the oxidation of the substrate, or the passage of the reducing equivalents down an oxidation-reduction chain such as the 'electron-transport' system of mitochondria or of the microsomal respiratory chain. This concept of 'electron-transfer' is seen simply if we consider the reduction of ferric to ferrous iron, i.e. $Fe^{3+} + e^{-} \rightarrow Fe^{2+}$.

Although the movement of electrons may be sufficient to explain the oxidation of iron-containing molecules such as cytochromes, it is known that dehydrogenation by enzymes which have NAD^{+} or $NADP^{+}$ as the coenzyme does involve the movement of hydrogen as well as of electrons, i.e.

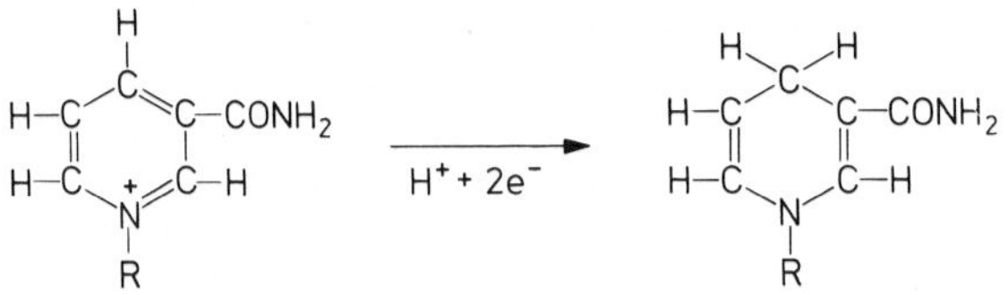

$$\xrightarrow[H^{+} + 2e^{-}]{}$$

Dixon and Webb (1964) considered that the electron-transfer concept could be misleading and preferred to speak of hydrogen-transfer. For simplicity, this terminology will be adhered to in this book; rigorously 'the reducing equivalents' should be read in place of 'hydrogen'.

III. Use of an Intermediate Hydrogen Acceptor

If neotetrazolium chloride is exposed, at a physiological pH, to NADPH, no reaction occurs over at least 30 min. Thus it follows that the reduced coenzyme cannot readily reduce the neotetrazolium chloride, so that any reaction which occurs in the presence of a tissue section will be due to the intermediacy of diaphorase activity in the section (as will be discussed in the next section). Consequently if we are to measure dehydrogenase activity directly, i. e. the amount of NADPH generated from $NADP^{+}$ by the removal of hydrogen from the substrate, we require to add an intermediate hydrogen acceptor which will remove hydrogen quantitatively from NADPH and transfer it, equally quantitatively, to the final hydrogen acceptor, neotetrazolium chloride. For many tissues, the most effective intermediate, in conventional biochemical as well as cytochemical studies (Altman 1972), is phenazine methosulphate (which has the endocrinologically unfortunate abbreviation of PMS). For other tissues, in which phenazine methosulphate may be inhibitory, menadione may fulfil the same role (Chayen et al. 1973c) even as an intermediate carrier from NADPH.

Other tetrazolium salts are reduced directly by NADH and by NADPH but the stoicheiometry of their reduction is not known. Even when they are used, the addition of phenazine methosulphate usually increases the intensity of the reaction. Furthermore the site in the various hydrogen-transport systems of cells at which they accept hydrogen (or reducing equivalents) is also uncertain (for a discussion of this question please see Altman 1972, 1976). Consequently the present author prefers to use neotetrazolium chloride which, on its own, appears to accept reducing equivalents from close to the most electro-positive (oxygen) end of the hydrogen-transport systems of cells, but which can be reduced stoicheiometrically by the reduced coenzymes through the intermediacy of phenazine methosulphate. Thus two measurements can be made: the first, with the intermediate hydrogen acceptor (phenazine methosulphate, or menadione) measures the activity of the primary dehydrogenase enzyme; the second, in the absence of any intermediate other than those provided by the cells themselves, gives an indication of the efficiency of the relevant hydrogen-transport chain.

IV. Validation

Altman (1967) showed that the colloid stabilizer, polyvinyl alcohol, did not affect the rate of oxidation of glucose-6-phosphate by a semi-purified preparation of glucose-

6-phosphate dehydrogenase, as assessed by the increase in the absorption at 340 nm of the NADPH generated in solution in the cuvette of a spectrophotometer. The activity/time graphs showed that the enzyme activity decayed sooner in the normal reaction medium than it did in the medium containing the stabilizer.

In a subsequent series of studies the activity of the two 'soluble' dehydrogenase enzymes of the hexose monophosphate pathway, namely glucose-6-phosphate and 6-phosphogluconate dehydrogenases, was assayed in samples of the liver of the same rat either by conventional homogenate biochemistry or by cytochemistry. The results, converted to absolute values (μmol of hydrogen/unit mass of tissue/unit time) were identical and reproducible (Altman 1972; Chayen 1978 b). Investigations of the activities and characteristics of mitochondrial oxidative enzymes yielded results in agreement with general biochemical findings (e. g. Butcher 1970; Altman 1972).

Thus it seems clear that the processes of chilling, sectioning, flash-drying and of reacting in the presence of a colloid stabilizer, retain the full activity of these oxidative enzymes. The reaction product, namely the precipitated coloured formazan, is quantitatively equivalent to the amount of reduced coenzyme normally measured in solution by conventional biochemical methods. But, because the formazan is precipitated at the site of its generation, it is possible to measure the activity in each selected cell, provided that a suitable microspectrophotometric method of measurement can be used (as will be discussed in Chap. 6). Consequently it has become possible to measure the biochemical activity of individual cells, knowing that the total biochemical activity of all the cells will indeed add up to the activity which could be measured, by more conventional techniques, in a sample of at least 1 million cells.

V. Function of Reducing Equivalents

A particular advantage of this type of cytochemical analysis (Fig. 5.4) is that the total amount of NADPH generated in unit time (in the presence of phenazine methosulphate) can be measured in one section, and the amount used by the microsomal respiratory pathway (measured in the absence of phenazine methosulphate) assessed in the next serial section. The latter has been termed Type I hydrogen. The difference between this, and the total NADPH generated (known as Type 2 hydrogen), gives an indication of how much NADPH is available for biosynthetic activity or for other mechanisms such as transhydrogenation reactions. These ideas have been discussed fully by Chayen et al. (1973 a, 1974 a) who showed their advantage in the analysis of the effects of various steroids. It was also found that in the zona reticularis, where the reducing equivalents are required for steroidogenesis and therefore for the cytochrome P-450 system, 96% of the very active generation of NADPH is of Type 1. In contrast, in rat adipose tissue, where NADPH is involved in the synthesis of fat, 96% is of Type 2 (Chayen et al. 1974 a).

VI. 'NADPH-diaphorase' System

The term NADPH diaphorase is applied, in cytochemistry, to any enzymatic oxidation-reduction system which reoxidizes the reduced coenzyme NADP at the

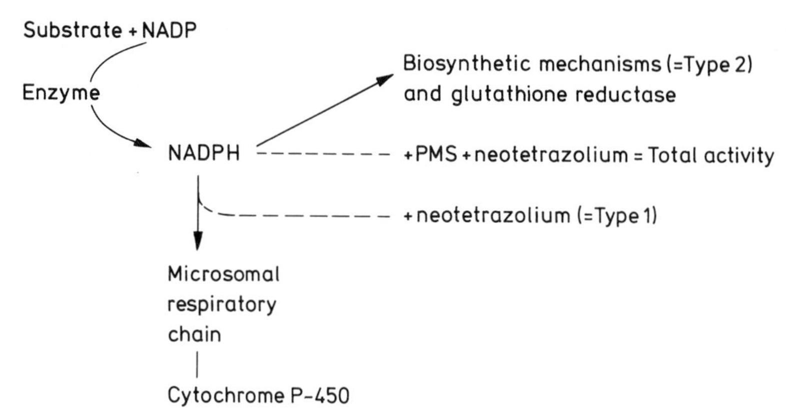

Fig. 5.4. NADPH can be used either for biosynthetic mechanisms or for transhydrogenation or for the microsomal respiratory pathway. The amount flowing down the last can be measured if neotetrazolium chloride is used alone. The amount available for other mechanisms is indicated by the difference between the total amount of NADPH generated, and the amount used in microsomal respiration (i. e. Total – Type 1 = Type 2)

expense of a tetrazolium salt which thereby becomes reduced to a coloured formazan. Frequently it is synonymous with the microsomal respiratory pathway (Chayen et al. 1973 a, 1974 a). It is measured as Type 1 NADPH hydrogen (above). The same general statement pertains to NADH diaphorase.

The reduced coenzyme NADPH is generated from its oxidized form ($NADP^+$) by three main dehydrogenase enzymes. Two, namely glucose-6-phosphate dehydrogenase and 6-phosphogluconate dehydrogenase, occur in the cytosol, are found in the supernatant fractions of homogenates and are the first steps of the hexose monophosphate pathway, or pentose shunt, which is of significance in cellular proliferation (Coulton 1977). The third, similarly present in the cytosol, is the NADP-dependent isocitrate dehydrogenase. All may provide NADPH for the microsomal respiratory pathway which terminates with a moiety known as cytochrome P-450. This system is the mixed function oxidase which, among other properties, hydroxylates many diverse types of molecules, including steroids (e. g. see Gillette et al. 1969). In the adrenal, this system may occur both in the cytosol and in the mitochondria (Simpson et al. 1969).

Phenobarbitone induces the microsomal respiratory pathway in the liver. In rats, this drug has been shown to increase Type 1 NADPH-hydrogen to the same extent as conventional biochemical methods have shown the microsomal respiratory pathway to be activated (Chayen et al. 1973 a). The particular point of interest in this system has been that the cytotoxicity which it produces through forming the toxic hydroxylated compounds of otherwise inert substances, causes predominantly centrilobular necrosis of the liver. Cytochemistry which, unlike conventional biochemistry, can relate chemical activity to histology, has shown that this system is most active in cells close to the central vein; it is least active periportally. Moreover, phenobarbitone induces this activity mainly in the centrilobular cells (Chayen et al. 1973 a; Chayen 1978 b). Thus cytochemistry has been able to explain this phenomenon of selective toxicity. Furthermore it has been able to measure the cytochrome P-450 by microdensitometry of the natural absorption characteristics which gave this name to this moiety (Altman et al. 1975; Gooding et al. 1978).

However, our main interest in this system, whether it is studied as the NADPH diaphorase, or Type 1 hydrogen, or the microsomal respiratory pathway, will be firstly as to whether it is involved in the hydroxylation of vitamin D and its derivatives and secondly in its involvement in thyroid metabolism.

C. Cytochemical Investigation of Lysosomal Function

I. General

From the pioneering and now classic studies of de Duve and his school, it became clear that lysosomes were cytoplasmic particles, of the size of small mitochondria, in which were contained the greater proportion of the hydrolytic enzymes of the cell, sequestered from the cytoplasm by a semi-permeable membrane (e. g. de Duve 1959, 1963, 1969). Thus the essence of these organelles was the latency of the enzymes within the organelles; they were not detectable biochemically until the membrane around the organelles had become disrupted. Thus the biochemist first measured the apparent enzyme activity of his lysosomal fraction; this gave him the ‘free’ activity of the hydrolytic enzyme being studied. He then disrupted the lysosomal membranes to disclose the full, or ‘total’, activity of the enzyme (Fig. 5.5). Among other methods of labilizing the lysosomal membrane, incubation at 37° C in an acetate buffer of acidic pH (e. g. pH 5.0) was effective.

The concept that enzymes could exist in a latent form, active but not demonstrably so because of the membrane which bounded the organelle, caused concern among certain biochemists. Yet it was a commonplace concept to cytochemists. Thus it was soon shown that, whereas the biochemist could detect only two states of the lysosomal membrane, namely either totally impermeable or totally permeable, the cytochemist could demonstrate an intermediate condition in which the lysosomal membrane was permeable to substrates but still retained the enzymes within the organelles (e. g. Bitensky 1963 a). Moreover the degree of permeability, or ‘fragility’, of the membrane could be shown to vary under pathological conditions and, in later studies, under the influence of pharmacologically active substances. (The cytochemistry of lysosomes has been fully reviewed by Bitensky et al. 1973; Bitensky and Chayen 1977; also see Chayen 1978 a).

In general the cytochemist has been more interested in the state of the lysosomal membranes than in the amount of enzymatic hydrolytic activity which is contained within them. The former reflects changes in the cellular physiology whereas the latter, which can also be measured cytochemically, indicates only increase in numbers of these organelles, or synthesis of the enzymes. More recently, with refined techniques, biochemists too are coming to study the functional state of the lysosomal membrane as an indicator of cellular physiology (e. g. Peters et al. 1975). But it will be obvious, on reflection, that the state of these membranes is likely to be severely altered by the processes of homogenization and differential centrifugation into a foreign medium so that it may not be astonishing if the biochemical assessment of lysosomal membrane function is rendered less sensitive because of these disruptive techniques (Allison 1968).

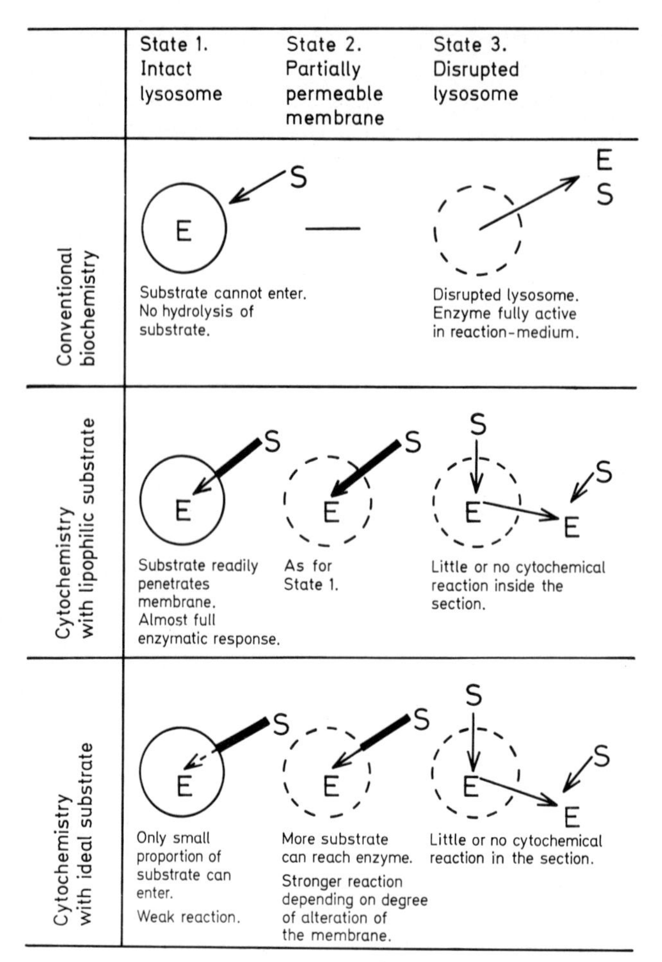

Fig. 5.5. The three possible states of lysosomal membranes; E, lysosomal enzyme; S, chromogenic substrate. *Top*, conventional biochemistry distinguishes between state 1 (intact lysosome) and state 3 (completely disrupted lysosome). *Middle*, a highly lipophilic substrate readily permeates through the lysosomal membrane so that it is virtually impossible to distinguish between states 1 and 2. *Bottom*, cytochemistry, done with a substrate that does not readily permeate through the intact lysosomal membrane, can disclose the three states of the membrane. The difference between low activity due to state 1 and that due to state 3 can be demonstrated by subjecting serial sections to acidic pretreatment: if it is due to the lysosomal membranes being intact, the pretreatment will cause increased activity (as in Fig. 5.7)

II. Cytochemical Investigation of Lysosomal Function

There are good cytochemical methods for demonstrating many of the lysosomal enzymes (Chayen et al. 1973 b, 1975; Bitensky and Chayen 1977). The drawback to many of these techniques is that the substrate is lipophilic so that it readily penetrates the lysosomal membrane. In consequence of this penetrability, it is impossible to measure changes in the permeability of the lysosomal membrane (Fig. 5.5). Only two of the substrates which have been tested up to now do not suffer from this drawback. The first is β-glycerophosphate, used as substrate for acid phosphatase. This is hydrophilic, and does not readily enter an intact, impermeable lysosome (state 1, Fig. 5.5). But with increasing exposure to acetate buffer at pH 5.0, at 37° C, the membranes become modified so that they will permit entry of this substrate. Consequently the state of the lysosomal membranes can be assessed by the time required, in acetate at pH 5.0 at 37° C, to produce a just appreciable reaction. This is the basis of Bitensky's (1963 a) fragility test for lysosomes. However, this reaction has not been favoured for more detailed quantitative study because the final reaction product is a small black granule of high optical extinction, which the present author did not consider to be ideal for microscopic densitometry. In spite of this, Aikman and Wills (1974) have measured such reactions in kinetic studies on lysosomes and have successfully elucidated the effect of ionizing radiation on lysosomal membranes as well as on intra-lysosomal acid phosphatase activity.

The second substrate is an amino acid naphthylamide, particularly leucine 2-naphthylamide. This does enter into intact lysosomes but at a relatively slow rate. The reaction is done in acetate buffer at pH 6.5; this is a compromise pH between that which may be optimal for the intralysosomal arylamidases or naphthylamidases (Tappel 1969) and that which is required for the simultaneous capture reaction where the naphthylamine, liberated by hydrolytic cleavage of the substrate, is coupled to a diazonium salt (tetrazotized dianisidine) to yield an insoluble azo dye (as discussed in Chap. 1).

When serial sections, containing relatively impermeable lysosomes (state 1 lysosomes) are reacted for increasing times in such a reaction medium, a response of the type shown in Fig. 5.6 is obtained. This implies that, initially, very little of the substrate (*S*) is reaching the enzyme. With longer incubation in acetate, even at this pH, the membrane becomes labilized and allows increasing concentrations of the substrate to enter the organelle and so be available to the enzyme. Over the region

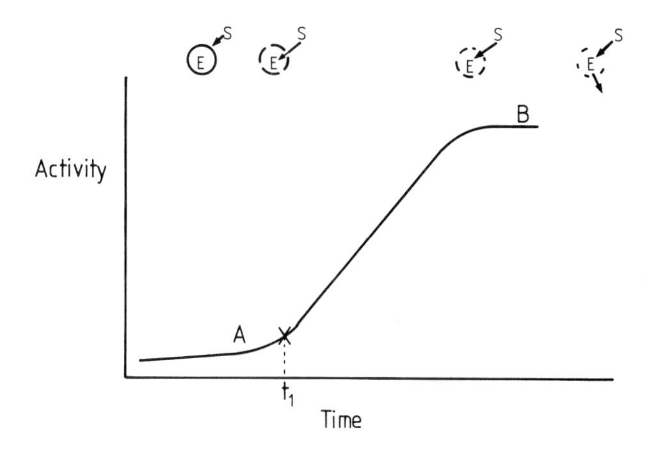

Fig. 5.6. Activity/time response of intact lysosomes. Over the region A to B, the curve increases and flattens out with time, very much as expected for enzyme activity measured in the absence of restraint. Up to time t_1 there is a lag period, caused by the relative impermeability of the lysosomal membranes. With increasing time, there is increasing labilization of the membranes. At longer times (after B) there may be loss of enzyme (as shown by the small diagrams of lysosomes at the top of the graph: as in Fig. 5.5), which also ensures that the activity within the lysosomes does not increase with time of reaction beyond this point

A-B the increase in activity is similar to that which would be obtained for an enzyme, free of membrane restraint, which is exposed to increasing concentrations of substrate, as in a normal substrate-activity graph as used in conventional enzymology.

From this graph (Fig. 5.6) we know the minimum time, t_1, which is required to obtain a significant response. We can now labilize the lysosomal membranes experimentally by pretreating the sections in acetate buffer at pH 5.0 at 37° C. Then, after increasing times of acidic pretreatment, all the sections are reacted in the chromogenic reaction medium for time t_1. The result (Fig. 5.7) is very similar to that found previously (Fig. 5.6) except that (1) the long lag period (up to A in Fig. 5.6) has been removed, and (2) there is a loss of enzymic activity when the pretreatment has been so extended that it has over-labilized the membranes and so allowed the escape of the arylamidases into the acidic pretreatment solution. Consequently there is a lower enzymic content (E′) to the lysosomes when they are finally exposed to the chromogenic reaction medium (for time t_1).

From such a graph (Fig. 5.7) we can derive the manifest, or 'freely available', activity of the intralysosomal naphthylamidases to this substrate (x) and the maximal, or total, activity (y) which these enzymes can display when the restraint of the membrane is minimal to the substrate, yet sufficient to retain the enzymes within the organelles. The amount of 'bound' or latent enzyme activity (y-x) as a proportion of the total, or maximal activity (i. e. (y-x) /y) is a measure of the function or impermeability of the lysosomal membranes in the original sample. It is often expressed as percentage bound, or percentage latent, activity [100 (y-x) /y].

It will be seen (Chap. 10) that the stimulation of the thyroid gland by TSH, or by long acting stimulators, involves a change in the functional properties of the lysosomal membranes of the responsive cells.

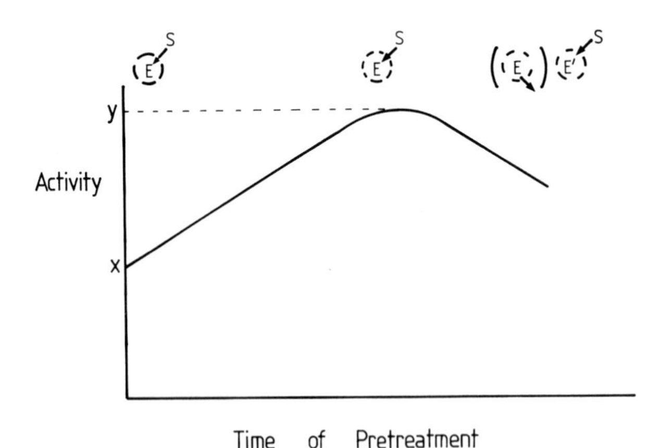

Fig. 5.7. The effect of acidic pretreatment. Sections are reacted, for the cytochemical reaction for the intralysosomal enzyme, for the minimum time (t_1 in Fig. 5.6) required to produce a measurable response (x) without acidic pretreatment. Increasing time of acidic pretreatment, prior to the cytochemical reaction for the same reaction time (t_1), increases the labilization of the lysosomal membranes, so allowing more substrate (S) to reach the intralysosomal enzyme (E) in unit time, and so increasing the observable activity (to y). With prolonged acidic pretreatment, enzyme is lost into the acidic medium so that there is less activity to give a coloured reaction product when the sections are immersed in the cytochemical reaction medium

Chapter 6

Microdensitometry

A. Development of Microspectrophotometry

I. Background

1. Artefact of the 'Average' Nucleus

If, for the moment, we exclude the work of Caspersson (as discussed in 6.A.I.2.), it can be said that microscopic photometry, or cytophotometry, developed around 1950 in response to two factors which occurred almost simultaneously. These were (1) the growth of cytochemistry, and particularly the concern to measure the amount of DNA in individual nuclei; and (2) the availability of reliable photomultipliers which were sufficiently small for use in conjunction with a microscope. The cytophotometry of the Feulgen reaction for DNA has been fully reviewed by several workers (particularly by Swift 1953; and by Leuchtenberger 1958). It soon found general favour by its ability to resolve a critical problem. Boivin et al. (1948; also Vendrely and Vendrely 1948) published biochemical analyses of the DNA content of nuclei of various tissues and showed that these were constantly twice the amount found in the sperm of the same animal. These results agreed with the view, developing even at that time, that DNA was related to the genic equipment of nuclei. However, some workers, using similar biochemical procedures, found that the amount of DNA 'per nucleus' could be 2.5 or even 3 times that in the sperm. Leuchtenberger et al. (1951) reinvestigated this problem both by cytophotometry and by conventional biochemical methods and demonstrated that in some tissues various degrees of polyploidy could confuse the biochemical estimations. For example, Leuchtenberger et al. (1951) obtained a value of 5.5 pg DNA per nucleus after extracting the DNA from nuclei of the rat kidney, but 8.2 pg per nucleus from nuclei of the liver. This higher value for liver nuclei was shown, by cytophotometry, to be due to three classes of polyploidy in liver nuclei: in a sample of 43 nuclei, 15 were diploid, 21 tetraploid and 7 octaploid. Cytochemically the diploid liver nuclei had the same amount of DNA as did the kidney nuclei, all of which were diploid. So it was shown that it can be misleading to isolate a large number of nuclei from a tissue; count them; extract the DNA and then divide the amount of extracted DNA by the number of nuclei to achieve a value for the DNA content per nucleus. Cytophotometry was then shown to be of especial value in measuring the DNA content of the individual nuclei in a proliferating population, where the amount of DNA per nucleus could vary from the 2c to the 4c value, depending on the time that had elapsed after the previous mitosis. Later it was used to measure the duration of the discrete period of DNA synthesis (the S-phase) that preceded mitosis (e.g. Deeley et al. 1957).

Just as different cells in a tissue may have nuclei which have different degrees of ploidy, certain cells or cell types may possess characteristic metabolic activities. It is the extension of this principle to cytochemistry generally that has allowed the development of the cytochemical bioassays in which the metabolic change, effected by the hormone, is measured specifically in the target cells.

2. Ultraviolet Microspectrophotometry and Potential Errors

The science of microspectrophotometry was developed by Caspersson during the 1930s (e. g. Caspersson 1940; and as reviewed by Caspersson 1947, 1950). Initially his interest was the measurement of nucleic acids and proteins in individual cells and, since both types of molecule have moieties which absorb in the shorter ultraviolet range of the spectrum (300–250 nm), his earlier work was concerned with ultraviolet microspectrophotometry.

Basically a microspectrophotometer for ultraviolet and visible light is a conventional spectrophotometer built around a microscope (Figs. 6.1, 6.2). The first critical difference is that in conventional spectrophotometry the material to be analysed is in solution, in a cuvette of standard light path (usually 1 cm), whereas in microspectrophotometry it is in a biological cell. Consequently a microscope system is employed to enlarge the specimen. This involves focussing convergent light on to the specimen instead of using collimated parallel light as in normal spectrophotometry. However, the increase in path length attendant on convergent light, which then diverges again before reaching the objective, is relatively unimportant (Walker 1958). Of greater moment is the overall path length when sections are used and this is the reason why it is essential to have sections of uniform thickness (as discussed in Chap. 5.A.5.).

Other potential errors inherent in microscopic spectrophotometry have been discussed at great length by various workers (as reviewed by Glick et al. 1951; Walker 1958). Goldstein (1970) considered the influence of stray light in the microscopical system and the degree to which out-of-focus effects might cause error (Goldstein 1971). In brief, none of these potential errors seems likely to cause errors of greater than about 3% in well-manufactured and properly handled instruments; they become critical if measurements more precise than this are to be made, as they are in very specialized studies. The serious problems in microspectrophotometry are

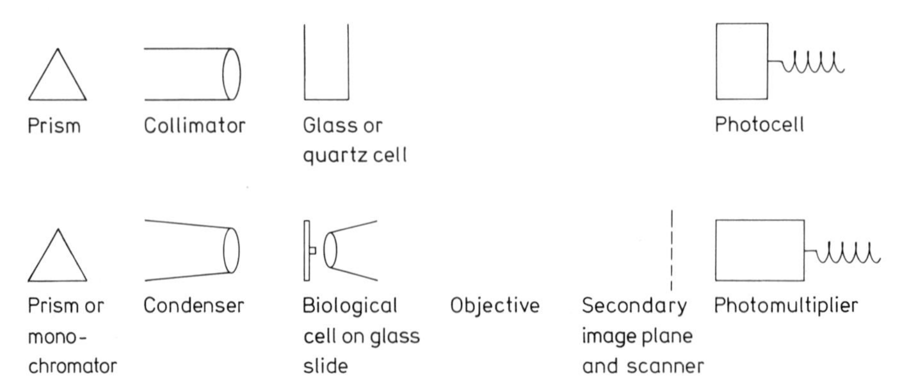

Fig. 6.1. Schemes for a conventional spectrophotometer (*top*) and for a microspectrophotometer (*bottom*)

Fig. 6.2. An ultraviolet and visible light microspectrophotometer (Zeiss UMSP. 1). At the extreme right is the xenon arc lamp which provides light to a conventional monochromator. Light is then passed either to a reference microscope (shown by a single draw-tube and eyepiece to the left of the main microscope) or to the specimen on the stage of the main microscope. The cathode-ray tube is to facilitate balancing the reference and measurement beams. The absorption (% transmission) is recorded on the chart on the extreme left, very much as it is in a conventional, double-beam, spectrophotometer

(1) diffraction, (2) scatter and (3) optical inhomogeneities. The last, which was relatively unimportant in ultraviolet studies on nucleic acids and proteins, will be discussed separately. The first two effects can be considered together since they involve either non-specific absorption or spurious gain of light.

3. Non-specific Absorption

Light can be lost from a section because it becomes scattered out of the optical system. Small particles, of the order of size of the wavelength of the light used, act almost as reflectors; the effect is analogous to the increase in light scattering caused by denaturation of proteins, or swelling of particles, as studied by conventional spectrophotometry. Caspersson (1940) examined this phenomenon in detail, as well as most potential sources of error. The loss of light caused by scattered light increases at lower wavelengths of the ultraviolet; the exact relationship is still not clear (whether it increases as the square or the fourth-power of the wavelength). In practice, this non-specific absorption of light, or loss of light, results in apparent absorption at a wavelength at which the chromophore does not itself absorb light. So, for example, when measuring the absorption of nucleic acids, the absorption at 310 nm is taken as being due to scatter only; a scatter curve is constructed to indicate how this amount of non-specific absorption will increase as the wavelength decreases to and through the region of the spectrum at which the nucleic acids absorb light. Then this scatter curve is subtracted from the absorption curve obtained in the specimen, so giving a measure of the specific absorption of the nucleic acids (and nucleoproteins). However, it is almost as precise to subtract the non-specific absorption at 310 nm from the absorption at 265 nm, the region of the spectrum at which nucleic acids in cells absorb maximally.

It will be seen that the same principles are applied in the cytochemical bioassay of ACTH. The specific reaction product for ascorbate absorbs maximally at 680 nm and minimally at 480 nm. It was found, by direct measurement of uncoloured sections, that the amount of light scattered from the steroids and lipids in the cells did not alter significantly over the range 480–680 nm, as would be expected at these relatively long wavelengths. Therefore the amount of the reaction product could be calculated by subtracting the extinction at the minimum from that at the maximum (i.e. $E_{680} - E_{480}$) since the absorption at 480 nm would have been zero had it not been for non-specific light loss.

4. Diffraction

The second major way by which loss of light, or gain of light, can distort measurements of absorption made through the microscope is by effects which can be grouped together as 'diffraction'. These can occur as follows. If we have a refractile object of very high extinction, light may become diffracted (bent) at its interface so that some light will now fall over the object. In general, the contribution made by such light would be small. But if the object is transmitting only 10% of the light incident on it, and if diffraction causes 10% of the light to be 'bent' over the object, then the error in absorption could be 100%. This is an extreme case, just to make the point that the extinction of each point in the specimen must not be high. An upper limit of 0.7 is probably safe. Of course it is unwise to measure high extinctions on conventional spectrophotometry; diffraction effects are additional hazards in microspectrophotometry. At low extinction values, where the overall extinction is about 0.3 (even though some regions may be higher and some lower), there is likely to be minimal possibility of error (Chayen and Denby 1968; Glick et al. 1951). In this context it may be noted that the medium in which the section (or cell) is mounted should be selected to match, as closely as possible, the refractive index of the cells to diminish both scatter and refractility problems. It is also advisable to use the largest diaphragm apertures, consistent with optimal performance, to reduce the effect of diffraction. Diffraction effects increase greatly when the light is almost parallel, which is why it is possible to see detail in an unstained section if the condenser diaphragm is 'stopped-down'. They are minimal when the light is strongly convergent, i.e. when lenses of high numerical aperture, and well-opened diaphragms, are used. However, if the field diaphragm is opened too widely, the light may not be accurately collected by the lenses and may reflect off the parts of the microscope to give 'glare'.

These problems, which may sound daunting, are eliminated routinely by using the correct procedure for setting-up the microscope system. They become problems only when workers make modifications to the mounting medium, or to the normal way of using a microscope, without understanding the bases of the procedures. They have been discussed in considerable detail in several specialist review articles such as those of Caspersson (1940), Walker (1958), Chayen and Denby (1968) and Goldstein (1970).

B. Microdensitometry

I. Basis of Measurement

1. Application of the Beer-Lambert Law

As has been discussed (in Chaps. 1 and 5) most cytochemical reactions, and at present all those used in the cytochemical bioassays, involve the precipitation of a coloured reaction product. This coloured precipitate is produced as the reaction occurs so that the amount of insoluble chromophore in each cell is a measure of how much of that specific biochemical activity is present in each individual cell. It remains only to measure the mass of the chromophore, in each cell.

In conventional biochemistry the amount of a coloured reaction product, in true solution, is measured by spectrophotometry. The basis of this form of measurement is the Beer-Lambert law, which can be applied to solutes provided that:

1) The solute is sufficiently diluted in the solvent that there are no intermolecular interactions between molecules of the solute. Such interactions give rise to secondary and even tertiary absorption maxima due to the formation of micelles of the solute (Michaelis 1947; Baker 1958). In histology and cytochemistry, such interactions are often seen as metachromasia. For example, although toluidine blue normally colours acidic groups blue, it can assume a red or purple hue if it forms micelles, particularly if such micelles result in suppression of ionization of the dye molecules (Chayen and Roberts 1955; Sylvén 1954).
2) The extinction of the solution is not inordinately high. Part of this proviso is the danger of interaction between the molecules of the chromophore, as just discussed. But another reason for this restriction is that accuracy declines at high extinction values even in a conventional spectrophotometer. The special reasons for this restriction in microdensitometry have been discussed by Glick et al. (1951).
3) There should be no interaction between the solute and the solvent or, if such interaction does occur, it should be taken into consideration. In cytochemistry the colour of the chromophore may change when it dissolves in fat or when it is bound to a more aqueous environment. But this can be seen; the absorption maximum and even the extinction coefficient of the aberrant form can be determined.

Lambert's law related the amount of light absorbed by different thicknesses of homogeneous coloured solids. It was converted by Beer to relate the absorbance, or extinction (E), of a chromophore in solution to the concentration (c) of that chromophore:

$$E = \log_{10} (I_o/I) = kct \qquad (1)$$

where I_o is the intensity of light falling on to the specimen;
I is the intensity of the light transmitted by the specimen;
t is the thickness of the specimen, or the light path, usually the dimension of the glass cell or cuvette used in the spectrophotometer;
and k is the extinction coefficient of that chromophore in that solvent.

Thus the concentration of the chromophore is given by the equation

$$c = E/kt \qquad (2)$$

The immediate point to be noted is that concentration is not related directly to percentage transmission, which would be $(I/I_o) \times 100$. Consequently we cannot measure the amount of a chromophore deposited in a cell by measuring the amount of light transmitted through that cell. Concentration is related logarithmically to the inverse of transmission.

2. Optical Inhomogeneity

In measuring biochemical activity per cell we want to determine the mass, not concentration, of the chromophore in each cell. Seeing that concentration is equal to the mass (M) per unit volume, and volume is equal to area (A) multiplied by thickness (t), we can say that

$$c = M/At \tag{3}$$

and, by substituting this in Eq. (2) we get:

$$M/At = E/kt$$

so that

$$M = E.\ A/k. \tag{4}$$

Equation (4) is adequate for measuring the mass of a chromophore dispersed homogeneously. But if it is precipitated in inhomogeneous precipitates, with clear spaces (100% transmission) in between, measurement of the whole area (A) by a photomultiplier can yield grossly erroneous results. For example, it has been calculated (Chayen 1978 b) that if a chromophore, which would have an extinction of 0.097 (i. e. transmitting 80% of the light), were precipitated so that it occupied half the field only, a simple measurement of the extinction of the whole field would be 11.3% in error. If the same chromophore occupied only 10% of the whole area, the error would be 58%. It was shown that it is theoretically possible to have two cells, one with four times the amount of chromophore as the other, and yet a simple measurement of extinction over both cells would be practically identical (Chayen 1978 b). Thus this optical inhomogeneity error could invalidate such measurements.

This optical inhomogeneity problem has been overcome by the development, by Deeley (1955), of scanning and integrating microdensitometry (or scanning and integrating microspectrophotometry). The essence of this system is to break up the area (A) of each cell, or group of cells, into very many (often a thousand or more) small regions of area a and to measure the extinction of each such region separately, integrating (or summating) all the measured extinctions (E_1, E_2, etc.) to give the total extinction of the selected area, A.

So that,

$$M = EA/k = a\,(E_1 + E_2 + E_3 + \ldots + E_n)/k \tag{5}$$

Provided that the diameter of a is smaller than the least optical inhomogeneity, the extinction of each region of area a will be measured accurately. However, if the diameter of a is greater than that of the optical inhomogeneities, each measurement will have the same inhomogeneity error as was discussed for the example when the chromophore fills only part of the measured field (also see Fig. 6.3).

In practice, when the chromophore is distributed in small dots, about 0.3–0.5 μm in diameter, as it is when lysosomal enzymatic reactions are done, the diameter of

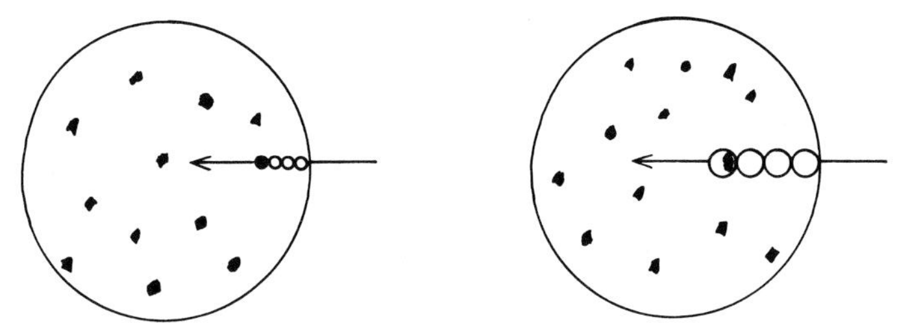

Fig. 6.3. Two microscope fields being scanned by a scanning spot. (The sequential position of the scanning spot, on one line of the scanning raster, is shown by the *open circles*.) On the left, the diameter of the scanning spot is equivalent to the size of the optical inhomogeneities (coloured, precipitated reaction product, shown as *dark irregularities*). On the right, the diameter of the scanning spot is about twice that of each of the optical inhomogeneities so that, when one clump of reaction product is encompassed by the spot, the measurement will be erroneous because half of the measured field transmits 100% of the light, whereas only part of the measured field contains absorbing matter

the scanning spot must be 0.25 μm. This is the limit of resolution of the light microscope so that no optical inhomogeneity of less than this size can be resolved as an inhomogeneity. With larger optical inhomogeneities, of diameter greater than 0.5 μm, it may be acceptable to use a scanning spot of 0.5 μm diameter.

The error caused by the inhomogeneous distribution of the precipitated chromophore will depend on the extinction of the precipitate. At optical densities (extinctions) of 0.2 to about 0.5 (Chayen and Denby 1968) or possibly 0.7 (Glick et al. 1951) the error is least and, provided that the extinction of the most intense regions is below about 0.5, the size of the scanning spot is less critical. However, when the chromophore is concentrated into very small spots, even though the mean extinction of the whole field (of area A) may be low, the extinction of the coloured regions can be at least 0.7 so that the accurate measurement of each of these regions will be critically dependent on the area (a) or diameter of the scanning spot (as in Fig. 6.3).

The theory and methods of microdensitometry have been discussed more fully by Chayen and Denby (1968), Chayen (1978 b) and by Goldstein (1970, 1971).

II. Use of a Scanning and Integrating Microdensitometer

1. Procedure

It has been emphasized that, for greatest accuracy, it is necessary that the diameter of the scanning spot should be capable of being made as small as 0.25 μm. The instruments having this capability, which have been commercially available for some years, are the Vickers M85 and M86 scanning and integrating microdensitometers; for simplicity, the operation of the former will be discussed.

The Vickers M85 microdensitometer (Fig. 6.4) is literally a spectrophotometer built around a conventional microscope. The latter has its own light source and field lens, supplying normal transmitted light into the conventional microscope. This allows the specimen to be inspected so that the histology can be ascertained and the responsive cell type can be identified. The magnification required is adjusted; this also controls the size of the scanning spot. An optical 'mask' is placed on the first cell or region to be measured. This mask directs the instrument to measure only what

Fig. 6.4. A scanning and integrating microdensitometer (Vickers M85). *a*, The lamp for the monochromator: *b*, the bandwidth control; *c*, the wavelength control; *d*, selector for different sizes of flying spot; *e*, housing for the measuring photomultiplier; *f*, specimen, mounted on a slide and viewed by normal microscopy. (Chayen 1978 a, p. 348)

lies within the mask; it is a way of 'dissecting-out', optically, the target cells from the mass of unresponsive cells which belong to different cell types.

The lamp for the spectrophotometer, the monochromator and all its controls, are housed above the microscope. They send a flying spot through the objective of the microscope (i. e. in reverse from the normal way of microscopy) across the mask, and each point is measured (Fig. 6.5). The light is passed through the condenser (which therefore acts as an objective in this respect) to the photomultiplier which lies in the base of the instrument.

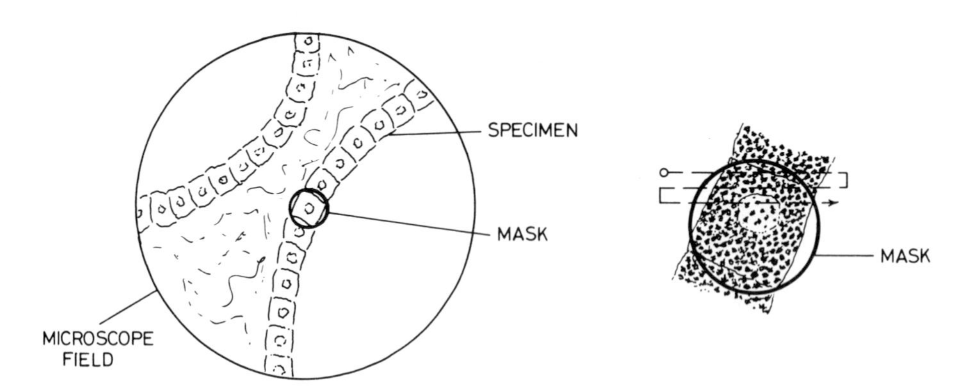

Fig. 6.5. Scanning and integrating microdensitometry. *left*, the histology of the specimen (a section through a thyroid) is ascertained by normal microscopy and the optical mask is placed on the first cell in which the absorption is to be measured. *Right*, the reaction product in this cell (shown at higher magnification) is scanned by a small flying spot. The absorption of each point in the selected cell is integrated to give the total absorption within the optical mask. The instrument ignores all light absorbed or transmitted outside the optical mask. (Chayen 1978 a, p. 349)

2. Mean Integrated Extinction

All the absorptions within the mask area are integrated and recorded by the digital meter. Similarly the instrument records the area occupied by coloured matter, or by the mask, depending on how it is instructed. The procedure for making these measurements will be discussed in Appendix 2; here we are concerned with what these measurements signify. The integrated absorption recorded on the digital meter is the summed relative absorption from all the points inspected within the mask. This 'relative absorption' can vary from one instrument to another, since it depends on the response of the particular photomultiplier used for making the measurements. Slightly different amounts of extra high tension applied to different photomultipliers, and indeed the individual responses of different photomultipliers, can produce very varied responses with dissimilar sensitivity of response. To produce a uniform response it is often preferable to convert these readings of integrated relative absorption into absolute units of mean integrated extinction. This is done very simply by producing a calibration graph relating the integrated relative absorption and absolute mean integrated extinction. This is achieved by using neutral density filters of known absolute extinction (which can be checked by placing the filters in a spectrophotometer). Each filter is placed so that it covers the microscope field. The amount of integrated relative absorption in the mask of the desired size, using the required microscopic magnification, is measured using a blank glass slide on the microscope stage, and each filter in turn (Fig. 6.6). Thus an integrated relative absorption, for a particular mask area, of a particular value (let us say 300 units) is shown to be produced by a field, of this size, where the *mean* absolute extinction over the whole field is, let us say, 0.3. The calibration graph

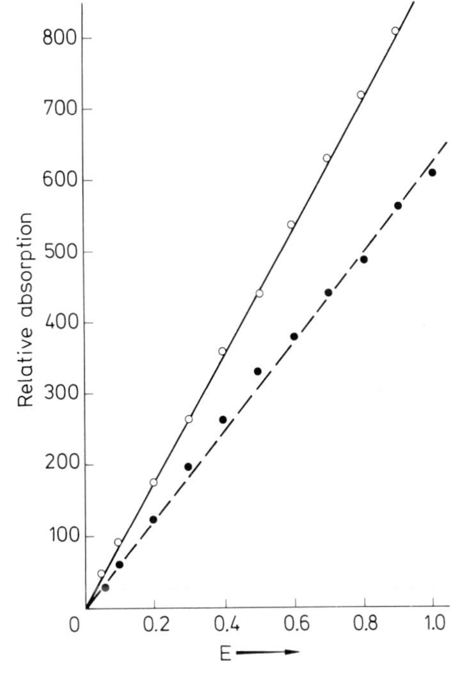

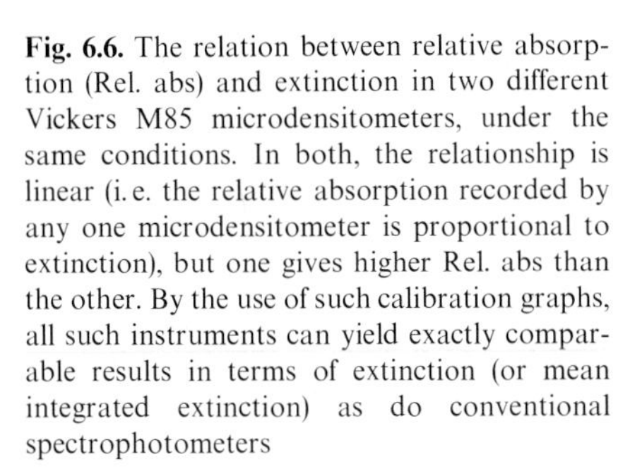
Fig. 6.6. The relation between relative absorption (Rel. abs) and extinction in two different Vickers M85 microdensitometers, under the same conditions. In both, the relationship is linear (i.e. the relative absorption recorded by any one microdensitometer is proportional to extinction), but one gives higher Rel. abs than the other. By the use of such calibration graphs, all such instruments can yield exactly comparable results in terms of extinction (or mean integrated extinction) as do conventional spectrophotometers

(Fig. 6.6) takes into account the individual response and sensitivity of this particular photomultiplier system. Consequently even though another instrument may record twice this integrated relative absorption (i.e. 600 units) for the same area, it will be found that its calibration graph will be steeper so that the mean integrated extinction will still be 0.3. Thus, irrespective of the instrument used, the results expressed as mean integrated extinction will be exactly comparable just as the extinction values recorded by one spectrophotometer will be identical to those measured by any other (provided they are properly calibrated and used correctly).

III. Validation of Microdensitometry

To measure the weight (M) of precipitated coloured reaction product, it is necessary to know the extinction coefficient of that chromophore in the solid state since $M = EA/k$. This is likely to be different, and sometimes very different, from the extinction coefficient in solution. But it is perfectly possible to measure the extinction coefficient of the precipitated chromophore, as has been done by Butcher and Altman (1973). By rearranging Eq. (4), namely $M = E.A/k$, they used the relationship $k = E.A/M$. In a sufficiently homogeneously coloured section they could determine E by microdensitometry. They could measure the area (A) of the section by planimetry of the enlarged image of the section, projected through a photographic enlarger. The mass (M) of the chromophore was measured by eluting all the chromophore from the section into a suitable solvent and measuring it in a spectrophotometer. For this measurement they used the extinction coefficient of the chromophore in this solvent, which could be determined directly (absorbance, or extinction per unit amount of chromophore added to the same solvent). This allowed them to determine the mass (M) of the chromophore which, spread out in the solid state in the section, produced a certain extinction (E) when measured by microdensitometry. Thus they could calculate the extinction coefficient of this chromophore when present as a solid precipitate. This allowed them to make absolute measurements of biochemical activity. Thus using a tetrazolium salt, which became reduced to a highly coloured insoluble formazan, they could measure dehydrogenase activity in terms of μmol of hydrogen removed from the substrate by the dehydrogenase enzyme in the cells within their sections, and compare the activities with more conventional biochemical assays of these enzymes in the same tissue. In the simplest case, namely for dehydrogenase enzymes normally found in the supernatant fraction of homogenates and therefore free from membrane restraint, they obtained the same values by microdensitometry as they did by conventional biochemical methods (as discussed by Chayen 1978b). For a mitochondrial enzyme, Butcher (1970) showed the same characteristics in sections as had been found by conventional biochemical procedures.

However, for most studies, it is not necessary to derive the extinction coefficient of the solid chromophore. It is enough that the mean integrated extinction/unit area is proportional to the mass (or amount) of the chromophore; it is not necessary to convert this proportionality into absolute units of mass unless comparison is to be made with results from some other technique. All the work in the cytochemical bioassays therefore depend on the relation $M \alpha E.A.$ since the constant of proportionality, the extinction coefficient of the solid chromophore, will be the same in all studies involving that particular chromophore.

The validity of microdensitometric measurements of cytochemical reactions has been discussed by Chayen et al. (1973 a, 1974 a) with particular relevance to NADP-dependent dehydrogenases and by Bitensky et al. (1973; Bitensky and Chayen 1977) in regard to lysosomal enzymes.

Chapter 7

The Segment Assay of the Adrenocorticotrophic Hormone

A. Rationale

I. Background

1. The Hormone

Until recently ACTH was regarded as a single, well-defined polypeptide of 39 amino acids. The structure of ACTH derived from the pituitary gland of several species, its physical chemistry and possible three-dimensional structure and conformational changes, have been reviewed by Schwyzer (1977; also see Ney et al. 1964). However, as shown by Orth and Nicholson (1977) it now seems to be only a 'central element in a wide array of biosynthetically-related polypeptides'. The potential precursor of all these hormones may be big ACTH (Rees 1977) or very big ACTH (Orth and Nicholson 1977) so that the lipotrophins (the large β_h-LPH and the γ_h-LPH), ACTH, the endorphins, α- and β-MSH, and the ACTH-like intermediate lobe peptide (CLIP) are all related. Thus α-enkephalin corresponds to β-LPH$^{61-65}$; β-endorphin to β-LPH$^{61-91}$; β-MSH which, in the human but not in other species, now seems to be an extraction artefact (Rees, 1977) corresponds to β-LPH$^{41-58}$ while α-MSH is equivalent to N-α-acetyl [ACTH$^{1-13}$]-NH$_2$; and CLIP is ACTH$^{18-39}$ (Orth and Nicholson 1977; Rees 1977). The possible relationships have been discussed by Smyth (1978) and are shown in Fig. 7.1.

Corticotrophin is a linear peptide composed of 39 amino acids. The species-specific amino acids reside in the carboxy-terminal two-thirds of the molecule which

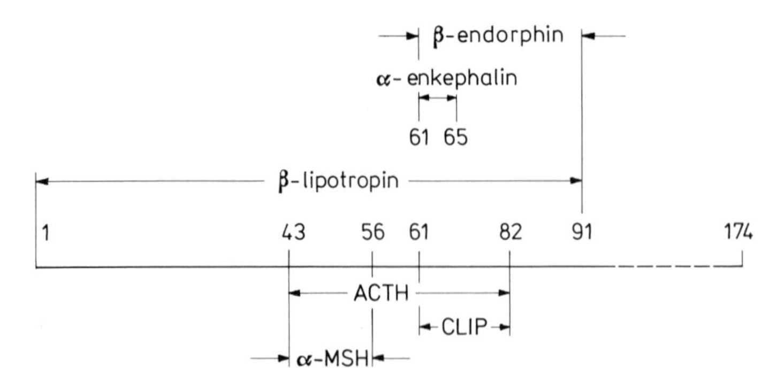

Fig. 7.1. Possible relationships and homologies between ACTH and other hormones which might be derived from a common precursor. (After Smyth 1978; Orth and Nicholson 1977)

appears to be involved in the stability of the molecule; in contrast, the amino-terminal part of the molecule contains the essential information for the cortico-trophic effects. Although slight biological effects have been observed with smaller fragments, at least the first 19 amino acids are required for high activity in vivo, with at least 50% of the total activity being observed with the peptide consisting of the 24-amino-terminal amino acids (Schwyzer 1977). In aqueous media ACTH is a flexible molecule which can adopt different conformations under particular conditions; being flexible means that it can adapt to different receptors with different binding constants and so produce a variety of effects (Schwyzer 1977). Corticotrophin has a helical conformation over two sections of its molecule, and a β-bend over other regions (Snell 1978).

The structure of the pig, sheep, bovine and human ACTH varies only in the region of the chain from amino acid 25 to 33 so that the critical first 24 amino acids are identical; the molecular weight of porcine ACTH is 4567 (Schulster 1974). Apart from the normal, or 'little' ACTH, other forms have been described although these may have little or even no biological activity. These include an 'intermediate' ACTH of mol wt. about 6000 daltons which had biological activity, big ACTH (mol wt. about 23,000) and very big ACTH (mol wt. about 31,000); it is possible that limited treatment with trypsin can convert the bigger forms into the smaller and that the big ACTH found in human tumours might be a glycoprotein (as discussed by Orth and Nicholson 1977). Gewirtz et al. (1974) isolated from tumours a big ACTH which had biological activity less than 4% of its immunoactivity, and which was degraded to little ACTH by treatment for only 10 s with trypsin, yielding full biological activity (also see Yalow 1974).

2. Biological Action of ACTH on the Adrenal Gland

The adrenal gland is affected by ACTH in a number of different ways (Schulster 1974). It increases the rate at which blood flows through the gland and it has the trophic effect, by which the hormone was known, of increasing the weight of the gland. There are some indications that the trophic effect might be capable of being dissociated from the steroidogenic activity (Ramachandran et al. 1977). The acute effects of ACTH acting on the adrenal gland include accelerated turnover of phosphate; stimulation of glycogenolysis and oxidation of glucose; decrease in the content of ascorbate, lipids and cholesterol; and the stimulation of the formation and increased content of corticosteroids (Schulster 1974). To these may be added a wide range of other biochemical changes, some obviously associated with steroidogenesis (e. g. see Kowal, 1970) and others, such as an increase in the rate at which proteins are turned-over, or even DNA-synthesis and the formation of a short-lived protein (Garren et al. 1971), which may or may not be directly involved in steroidogenesis (Koritz et al. 1977).

As do so many hormones, ACTH stimulates adenylate cyclase activity. However, the function of the cAMP produced is uncertain. Hudson and McMartin (1975) have advanced powerful evidence against cAMP playing a direct role in steroido-genesis in the adrenal cortex. Moyle et al. (1976) found that a derivative of ACTH produced full steroidogenesis while inhibiting the adenylate cyclase activity and leaving that of the histone kinase unaffected. On the other hand, dibutyryl cAMP has been shown to enhance steroidogenesis in adrenal cortical cells in various systems (as discussed by Schulster 1974).

Another important enzymatic system of the adrenal cortex stimulated by ACTH is ornithine decarboxylase and its associated pathway for the biogenesis of polyamines (Byus and Russell 1975; Levine et al. 1975). It is becoming increasingly obvious that the stimulation of this enzyme, followed by the synthesis of polyamines, is as regular a sequel of the attachment of a polypeptide hormone to its target cell as is the stimulation of adenylate cyclase. Indeed it is possible that there is a direct relationship between the two events. Too little has been done on this relationship with regard to the action of ACTH; further discussion of the possible function of polyamines must await consideration of the mode of action of TSH (Chap. 10).

II. Ascorbate Depletion and Hydroxylation

A series of studies by Sayers and his colleagues showed that exogenously administered ACTH, or conditions which would cause increased endogenous production of the hormone, caused a loss of ascorbic acid from the adrenal glands. This work culminated in the much used in vivo bioassay of ACTH based on this phenomenon (Sayers et al. 1948). By microchemical procedures, with histological monitoring of the preparations, Bahn and Glick (1954) repeated these findings and showed that the main regions affected were the outer zone of the zona fasciculata and the zona reticularis, the response in the latter (in monkeys and rats) being at least as great as that found in the narrow region of the outer zona fasciculata. The other regions of the zona fasciculata were less influenced by ACTH and had lower normal concentrations of ascorbate.

The function of this loss of ascorbate is not obvious even at present. The clearest evidence for its direct role in steroidogenesis comes from the work of Hodges and Hotston (1970) who showed that, once guinea-pigs had been rendered fully scorbutic, their production of cortisol and corticosterone became maximal, that is as great as any concentration of ACTH could induce in the non-scorbutic animals. Thus apparently the presence of ascorbate in the adrenal cortex inhibits the production of the end product of this pathway of steroidogenesis (Fig. 7.2).

A possible function of ascorbic acid in microsomal electron transport and in hydroxylation mechanisms in the adrenal gland was suggested by Staudinger et al. (1961). Since that time the basic form of the microsomal respiratory pathway and its peculiarities in the adrenal gland have been elucidated. It will be seen (Fig. 7.2) that the biosynthesis of cortisol, or corticosterone, from cholesterol involves several hydroxylations in some, or perhaps all, of which this pathway is implicated. Thus this pathway is of obvious importance in steroidogenesis and, by implication, in the mode of action of ACTH. The microsomal respiratory pathway, terminating in cytochrome P-450 (Fig. 7.3), in the adrenal gland and its involvement in these hydroxylations have been reviewed by Simpson et al. (1969). As these authors pointed out, it seems that during the transformation of a molecule of cholesterol to one of cortisol, the molecule apparently gets shuttled back and forth between the mitochondria and the endoplasmic reticulum. Why this should have to be is not understood, nor are many other facets of this apparently complex chain of interacting events. There is some evidence (McKerns 1966) that the key enzyme in supplying reducing equivalents (NADPH) for these processes might be glucose-6-phosphate dehydrogenase. Whether this is so or not (since no mechanism has yet

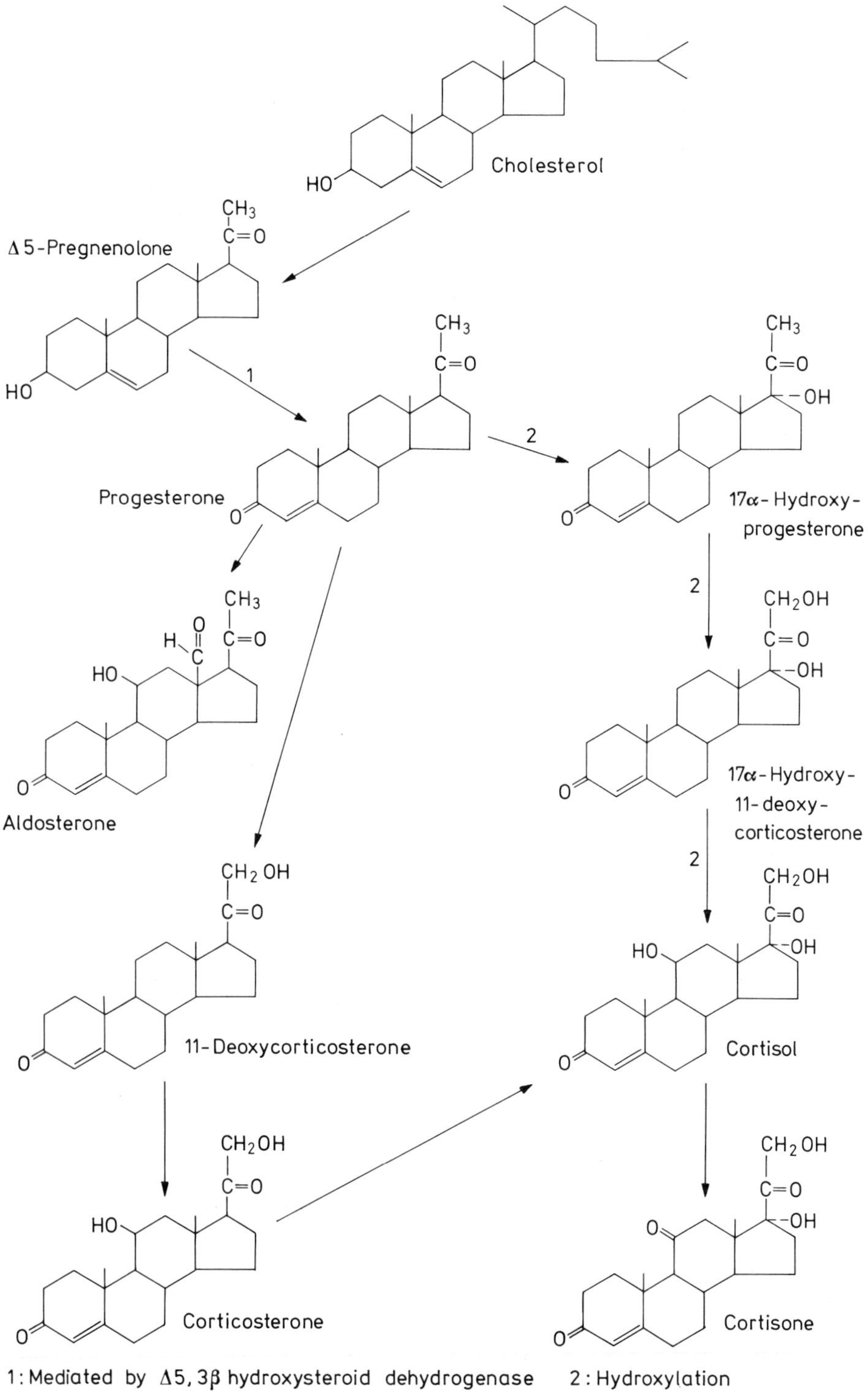

Cholesterol

Δ5-Pregnenolone

Progesterone

17α-Hydroxy-progesterone

Aldosterone

17α-Hydroxy-11-deoxy-corticosterone

11-Deoxycorticosterone

Cortisol

Corticosterone

Cortisone

1: Mediated by Δ5, 3β hydroxysteroid dehydrogenase 2: Hydroxylation

Fig 7.2. Pathways of steroidogenesis

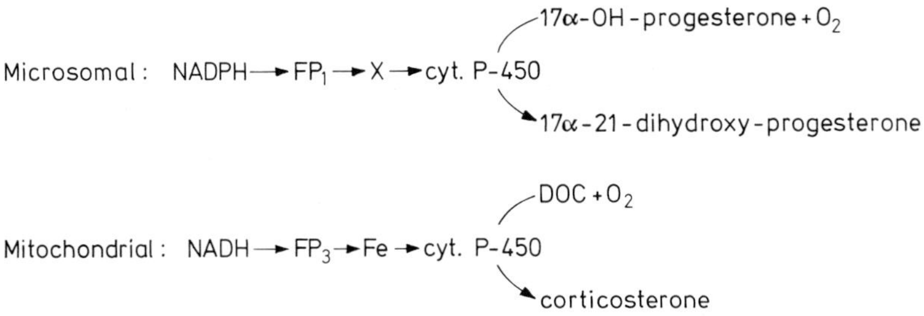

Microsomal: NADPH → FP_1 → X → cyt. P-450 → 17α-OH-progesterone + O_2 / 17α-21-dihydroxy-progesterone

Mitochondrial: NADH → FP_3 → Fe → cyt. P-450 → DOC + O_2 / corticosterone

where FP = flavoprotein
Fe = non-haem iron protein called adrenodoxin
cyt. P-450 = cytochrome P-450

Fig. 7.3. Microsomal and mitochondrial pathways involved in steroidogenesis. Only two examples of steroid hydroxylation are shown

been described by which NADPH itself can enter mitochondria), two requirements of this hydroxylating system are clear (Simpson et al. 1969). These are (1) a supply of reducing equivalents, preferably from NADPH (Ryan and Engel 1957) but possibly from NADH and transhydrogenating mechanisms, and (2) molecular oxygen (Hayano et al. 1955). Since ascorbate competes avidly for molecular oxygen, it has been suggested (Chayen et al. 1976) that the inhibitory influence of ascorbate in steroidogenesis, as demonstrated by Hodges and Hotston (1970), could be due to its competition against cytochrome P-450 for molecular oxygen. The presence of cytochrome P-450 in the adrenal cortex and the effect of metyrapone in inhibiting 11β-hydroxylation by this cytochrome P-450 system have been discussed by Wilson et al. (1968).

III. Hypotheses Concerning the Mode of Action of ACTH

Cholesterol esters appear to be stored in the cells of the zona fasciculata. ACTH mobilizes these stores (although it may also stimulate the new synthesis of cholesterol from acetate, according to Kowal 1970). The first steps involve the cleavage of the ester and the action of Δ^5, 3β-steroid dehydrogenase to convert pregnenolone to progesterone. This dehydrogenase is present in the cells both of the zona fasciculata and of the zona reticularis and its activity is enhanced even by low concentrations of ACTH as tested in a cytochemical bioassay system (Loveridge and Robertson 1978). There are many indications that the cells of the zona reticularis are involved in the final stages of the production of cortisol (e.g. Symington 1969). One of these indications is the fact that the phenomenon of ascorbate depletion is strongly marked in these cells. Another is the observation that steroids can be visualized by direct polarization microscopy in the cells of the zona fasciculata in guinea-pigs injected with dexamethasone to inhibit endogenous ACTH, but in the zona reticularis when ACTH is administered.

Thus it is suggested that the production of cortisol (in the guinea-pig) involves not only co-operation between the hydroxylating mechanisms of the mitochondria and endoplasmic reticulum, as just discussed, but also co-operation between cells of the two major regions of the adrenal cortex. This might explain some of the remarkable sensitivity shown by the cytochemical bioassay of ACTH, in which the cells of the different zones are left in their spatial relationships one to the other.

In our earlier studies we, like others, investigated the effect of ACTH on the supply of reducing equivalents from the activity of glucose-6-phosphate dehydrogenase (e. g. Chayen et al. 1976). Given optimal concentrations of reactants, this enzyme was remarkably active, particularly in the zona reticularis and the zona fasciculata interna, and ACTH made only a slight impression on this activity. But the intriguing feature was that almost all the reducing equivalents were apparently linked to what we termed the Type I hydrogen pathway (Chayen et al. 1974 a), i. e. it was not available as NADPH for biosynthetic mechanisms but was apparently associated with hydroxylating systems. Thus the reducing equivalents were apparently always available, and in plentiful supply, provided that the conditions were correct. The only condition which seemed to be limiting was the supply of molecular oxygen, which could be controlled by the intracellular concentration of ascorbate. And it was this condition which was remarkably sensitive to ACTH. Chayen et al. (1976) therefore suggested that the role of ACTH, in this context at least, was to deplete the intracellular concentration of ascorbate so as to allow molecular oxygen to become available to the cytochrome P-450 in these cells, and so allow it to utilize the reducing equivalents which were potentially available to it (Fig. 7.4).

B. Cytochemical Segment Assay of ACTH

I. Introduction

This assay was the first cytochemical bioassay to be developed (Chayen et al. 1971, 1972; Daly et al. 1972 a). Initially it was developed to help the investigation of the hypothalamic-pituitary-adrenal function in patients with rheumatoid arthritis who had been treated with either glucocorticoid or ACTH therapy (e. g. Daly et al. 1974 b). It proved of value in other physiological and clinical studies (as reviewed by Daly et al. 1974 a). It was a research tool which sufficed our needs in these particular investigations. However, its success brought demands for a greatly increased through-put so that the assay could be used more routinely, as will be discussed in relation to the section assay (below). Although the cytochemical section assay has now largely superseded the segment assay (as discussed by Daly et al. 1977), the

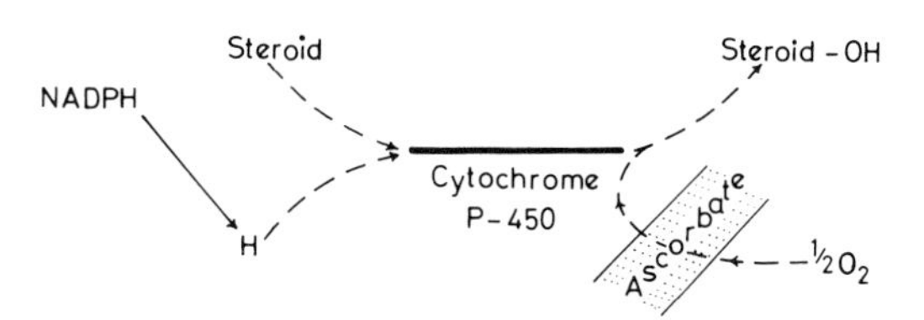

Fig. 7.4. Hypothetical scheme relating ascorbate depletion to the level of cytochrome P-450 activity. Steroid hydroxylation requires a supply of hydrogen from NADPH, and atmospheric oxygen. Ascorbate reacts preferentially with the latter and so may act as a barrier to its access to cytochrome P-450. Depletion of ascorbate would restore the activity of the cytochrome P-450. (Chayen et al. 1976, p. 45)

segment assay procedure is more adaptable so that there are still many research investigations for which it is preferred.

II. Procedure

1. Target Tissue

The animal used for this assay is the guinea-pig, of the Hartley strain, of a weight about 500 g, the sex being immaterial. It is killed by asphyxiation in nitrogen. It is important not to use ether, or some similar material, which can affect cell membranes and cellular lipids (Gahan 1962). The two adrenal glands are removed, trimmed quickly of fat and connective tissue, and each is cut into three segments. These are best cut as shown in Fig. 7.5.

2. Apparatus

The culture apparatus (Chap. 4, Figs. 4.1, 4.2) consists of an outer cylindrical pot (e.g. 5 cm diameter and 9 cm tall), made of Perspex or a strong polyethylene material, fitted with a well-sealing lid in which inlet and outlet tubes are inserted by gas-tight seals. Inside this culture pot is placed a vitreosil dish in which stands a small table (2.5 × 2.5 cm) made of expanded stainless steel mesh. This table is covered with a small square of defatted lens tissue, prepared by washing microscope lens tissue repeatedly with ether and then allowing it to dry; it is then washed in three changes of distilled water and dried. The vitreosil dish, metal mesh table and lens tissue are sterilized. The culture pot is sterilized only by being washed thoroughly and exposed to ultraviolet light. (See Appendix 1).

3. Culture Medium and Procedure

Each segment of the adrenal gland is placed on the lens tissue on the table in one vitreosil dish and the culture medium is pipetted into the vitreosil dish up to the level of the top of the table so that the segment is sited on moist lens tissue above the level of the medium. It is essential that the medium does not reach the segment directly so that the cells are not immersed in fluid but are exposed to the atmosphere. Each vitreosil dish is placed in one culture pot, the lids are sealed with Lanolin, and the atmosphere within the pots is replaced by one containing 95% oxygen: 5% carbon dioxide by means of the gas inlet and outlet tubes. The gas is passed for only about 5 min, depending on the volume of the culture pot and the rate of flow of the gas. The culture pots, the gas cylinder, the wash-bottle through which the gas is passed, to moisten it before it enters the culture pots, and the tubes carrying the gas (Fig. 4.2, Chap. 4) should all be in a room which can be maintained at 37° C (35°–37° C) so that the gas will be at the required temperature. The function of the gas mixture is (1)

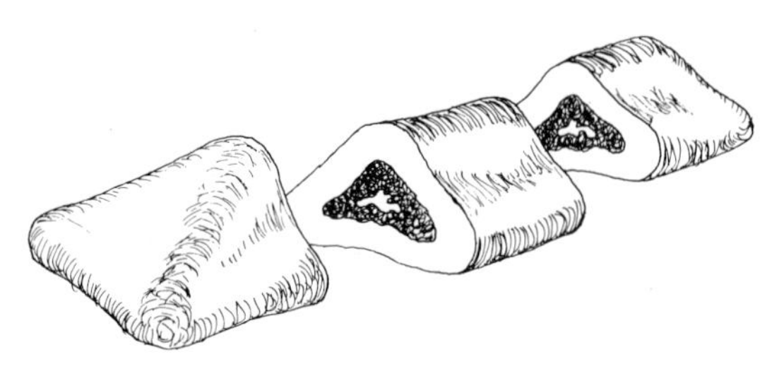

Fig. 7.5. The best way of cutting an adrenal gland into three segments, all containing the zona reticularis

to provide oxygen to the tissue and (2) to provide sufficient carbon dioxide to form a buffer system with the bicarbonate in the culture medium. This medium, Trowell's T8 medium, consists of a balanced composition of amino acids and salts, and bicarbonate. It includes phenol red as an indicator of the pH which, for these segments, should be pH 7.6. Even small variations (0.2 of a pH unit) can cause marked alteration in the sensitivity with which the cells respond to the hormone (Chayen et al. 1972 a). Before use, ascorbate is added to the culture medium to a concentration of 10^{-3} M (Chayen et al. 1972 a, 1976).

The segments are left at 37° C in these sealed culture pots for 5 h.

4. Function of the Maintenance Culture

This period of maintenance culture has a number of purposes. Firstly it removes the target tissue from the hormone endogenous to the guinea-pig. This is perhaps especially important in this assay because the stress of death may be expected to cause a marked rise in the level of ACTH circulating in the animal. But it applies to all the cytochemical bioassays in that greatest sensitivity will be obtained when the biochemical activity, stimulated by the endogenous hormone, has been allowed to revert to its basal, unstimulated level. Secondly the 5-h period allows the target tissue to recover from the trauma of excision and handling. But it must be realized that the tissue must be put into culture expeditiously: the longer the period between the death of the animal and the placing of the segments in maintenance culture, the less sensitive will be the response that the hormone can evoke in the target cells. Thus it is useless to do an exquisite dissection of the target organ if that procedure takes so long that the cells become unresponsive. The time taken should not exceed about 12 min. But even with the most dextrous handling, tissues show a time dependent biochemical response to excision and handling (Chayen et al. 1966; also see Chap. 4); 5 h are normally sufficient for them to recover from this effect. Thirdly the period of maintenance culture allows the tissues to replenish any ascorbate which they may have lost (Chayen et al. 1976), possibly as a consequence of the release of ACTH on death.

5. Exposure to the Hormone

At the end of the 5-h period the lids are removed and the culture medium is drawn out of the vitreosil dishes by means of a pipette. It is replaced by T8 medium with 10^{-3} M ascorbate containing the preparation of ACTH. Each of four segments receives a particular concentration of a standard preparation of ACTH, such as the 3rd International Working Preparation, or the pure α^{1-24} ACTH, at a concentration of either 5 or 0.5, or 0.05 or 0.005 pg/ml (Appendix 4). For each of the other two segments the T8-ascorbate medium contains the plasma at a concentration of either 1/100 or 1/1000. The optimal time of exposure to the hormone, in our hands (Chayen et al. 1974 b), is 4 min (but this time could possibly vary slightly and it may be helpful if it is measured in each laboratory doing this assay). At the end of this time the segments are chilled to −70° C in *n*-hexane and stored at −70° C until they are to be sectioned.

It must be emphasized that this assay, like all the other cytochemical bioassays, is done on plasma, not on serum. This is because the bioactivity of polypeptide hormones decreases rapidly in blood left at room temperature as it must be during the separation of serum.

6. Cytochemical Reaction

The tissue must be sectioned within a few days of its being chilled. Sections are cut at 12 μm in a Bright's cryostat fitted with an automatic cutting device to ensure even thickness of sections. The variation between serial sections should not be greater than ±5% (Chayen et al. 1972a; Daly et al. 1974c). The sections are reacted for reducing equivalents by immersing them for a total period of 12 min in three baths of freshly prepared Chèvremont-Frederic (1943) ferricyanide-ferric chloride reagent (Chayen et al. 1973b).

This reagent (Alaghband-Zadeh et al. 1972; Daly et al. 1974c) contains a certain proportion of potassium ferricyanide and ferric chloride (Appendix 3). Ferricyanide is reduced by many reducing agents and the ferrocyanide so produced links to ferric ions to form ferric ferrocyanide or Prussian (or Berlin) blue, which is an insoluble chromophore of high extinction coefficient. The proportions of ferricyanide to ferric ions are those which tend to respond more readily to ascorbate than to sulphydryl groups (Loveridge et al. 1975). However, it is by no means certain that ascorbate is the only reducing agent which contributes to this reaction in this procedure. Consequently in its earlier years this cytochemical bioassay was known as the redox bioassay. For all that, it has been shown that if a different and more specific reagent for ascorbate is used, namely an acidified solution of silver nitrate (Chayen 1953), the same results are obtained albeit more slowly (Loveridge et al. 1975).

The absorption spectrum of the Prussian blue deposited in the cells of the zona reticularis has been measured by means of a Vickers M85 microdensitometer (Fig. 7.6). The absorption maximum is at 680 nm and the minimum at 480 nm. However, at 480 nm there is still considerable apparent absorption, probably caused by the light scattered by the steroids and other lipids. Normally it is possible to match the refractive index of the tissue fairly closely by the refractive index of the medium in which the reacted sections are mounted. But in this tissue it was not found possible to match simultaneously matter of high refractive index, such as steroids, and the lower refractive index of the cellular matter. Thus the extinction, or

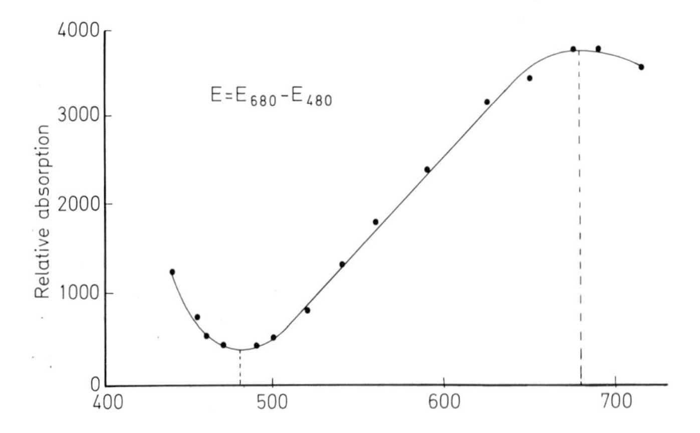

Fig. 7.6. Absorption curve of the Prussian blue reaction measured by microdensitometry in the cells of the zona reticularis. The absorption at 480 nm, at the point of minimal absorption, is due to 'scatter' in the cells. (Chayen et al. 1976, p. 47)

absorbance, at the wavelength of maximal absorption (E_{680}) is measured, as is that at the minimal absorption (E_{480}). Since it was found that the amount of scatter due to the tissue was relatively unaffected by the wavelength of the incident light, the absorption due to the specific chromophore can be obtained by the expression $E_{680} - E_{480}$.

7. Measurement

Prussian blue is deposited in the zona reticularis in such a manner that relatively large regions have the same extinction, i.e. the optical inhomogeneity is quite coarse. Moreover the intensity of each region is low (less than an extinction of 0.5 and may be less than 0.3). Consequently it is permissible to use an objective of low magnification ($\times$ 20 or $\times$ 10) which means that the size of the scanning spot will be correspondingly large (e. g. 1 or 2 μm diameter). The size of the mask may be large, but not so large as to include a variable number of unreacted spaces between blocks of cells of this zone. A suitable combination would be as follows: mask diameter of between 90 and 160 μm depending on the specimen; objective $\times$ 10 (or occasionally $\times$ 20); spot size 1 giving a diameter in the plane of the section of 2 μm.

Ten measurements are made in each duplicate section for each point in the assay, making 20 measurements for each point. It is normal to correct the measured values, which are relative absorption values, to the mean integrated extinction (as discussed in Chap. 6). Thus each point in the assay is the mean of 20 mean integrated extinction values. Each point can be expressed as the mean $\pm$ standard deviation, or mean $\pm$ standard error of the mean (SEM), or as the mean with the actual values for each slide to show the variation between the duplicate sections. Since the response of the cells is recorded by subtracting one reading from another ($E_{680} - E_{480}$), so involving the standard deviation of two populations, the last way of presenting the results is probably preferable. This is because the rapid method of making these measurements is to make ten readings at 480 nm, having first adjusted the microdensitometer to '100%' transmission for this wavelength, and then to adjust to 100% transmission at 680 nm for the next ten readings in the zona reticularis. The mean value of the first set of readings is subtracted from the mean of the second (to give the value of $E_{680} - E_{480}$). Since most of the zona reticularis present in the section will be covered by each set of ten readings this is a permissible manoeuvre. However, if a precise mean with either standard deviation or SEM is required, each field will have to be measured first at one and then at the other wavelength. Since this means moving to a clear background to adjust the conditions to give 100% transmission for each wavelength, this will delay the assay considerably. Alternatively some workers find that the E_{480} is so constant that they decline to measure it and use the E_{680} value alone. Personally I do not recommend this procedure except, possibly, for the most expert workers.

III. Results and Validation

1. The 4 + 2 Assay: Precision

The activity ($E_{680} - E_{480}$) of each sample (mean and duplicate values) plotted against the log concentration of the hormone (Fig. 3.3., Chap. 3) shows a negative linear correlation of shallow slope (b) but with little variance (s_{yx}). Thus although b is small (about 3.5: Daly et al. 1974c), since $s_{y \cdot x}$ is also small, the index of precision (λ), based on the readings shown in Fig. 3.3, is also relatively small (since $\lambda = s_{y \cdot x}/b$). As reported by Daly et al. (1974 c) the index of precision was between 0.05 and 0.09 and this has been the experience of ourselves and other laboratories since that time. Correspondingly (see Chap. 3) the values for the two dilutions of a plasma have been in agreement to $\pm 15\%$. This is a test of whether or not the activity of the plasma sample parallels that of the standard. On a number of occasions five serial sections have been measured for each point of the assay. The coefficient of variance for each set of five was always less than $\pm 5\%$. Although these are not independent samples, the fiducial limits (at $P = 0.95$) calculated from them, in 2 + 2 assays (Gaddum 1953; British Pharmacopoeia 1963) have ranged from 87%–115% down to 98%–102% (Daly et al. 1974 c). The former figure is the more usual.

2. Sensitivity

The least concentration of the hormone which can be measured above the effect of the vehicle alone (fresh T8 medium with ascorbate) is about 5 fg/ml (Daly et al. 1974 c). Holdaway et al. (1974 a) recorded a value of 10 fg/ml. For measuring circulating levels of the hormone the final volume of the sample, placed in the vitreosil dish, must be 10 ml. Thus at a dilution of 1/10, it should be possible to measure a circulating level of 50 fg/ml. In practice, using such a dilution, Rees et al. (1973 b) measured the circulating level as 34 fg/ml in a volunteer to whom 100 mg of cortisol had been administered to suppress the secretion of ACTH.

3. Reproducibility

As has been noted above, the reproducibility between serial sections is $\pm 5\%$. When three segments of the adrenal glands from one guinea-pig were exposed individually to one concentration of the hormone (0.27 pg/ml) the results agreed to $\pm 3\%$ (Daly et al. 1974 c; a similar study gave $\pm 4\%$: Chayen et al. 1972 a). Interassay variation, i. e. the variation found in assaying the same sample repeatedly, has not been rigorously examined. In one study (Chayen et al. 1972 a) the same plasma assayed on two successive days gave identical results; later a decline in value was noted, but this could have been due to deterioration of the biological activity with repeated thawing and freezing.

4. Accuracy and Specificity

In recovery studies aliquots of a sample of plasma were first assayed to determine the content of ACTH in that plasma. To each millilitre of two other aliquots were added 250 pg of a preparation of ACTH (101 IU/mg). These reinforced samples had to be diluted 1/10,000 and gave the recovery of the added ACTH (final value – initial value of plasma, as a percentage of the added concentration of ACTH) of 88% and 94% respectively (Chayen et al. 1972 a).

Luteinizing hormone, which can cause ascorbate depletion in the ovary, synthetic αMSH (MSH = melanocyte stimulating hormone), synthetic α^{18-39} ACTH and ovine prolactin gave no response in this assay even at molar equivalents of the peptides of 10^4 times and greater. Synthetic bovine βMSH, which has some amino acid sequences in common with ACTH as discussed above, cross-reacted less than 0.01% on a molar basis (Holdaway et al. 1974 b). In a subject given 100 mg of cortisol intravenously the endogenous circulating ACTH activity declined from 43 pg/ml to 34 fg/ml over a period of 4 h, with a half-time of endogenous ACTH-like activity, calculated from the first exponential part of the graph, of 10.4 min (Rees et al. 1973 b; Holdaway et al. 1974 b). This half-time is in fair agreement with that found by Meakin et al. (1959). Thus the ACTH-like activity measured by this cytochemical bioassay appeared to behave in a way which was identical to that of ACTH; this will be elaborated further in the physiological studies done with this assay.

As a further test of specificity Holdaway et al. (1974 b) added a specific antibody to ACTH (a 1/400 dilution of an antiserum to the N-terminal region of ACTH). They used a precipitating antibody against γ-globulin to precipitate the antibody-ACTH complex. With suitable controls they found that the specific antibody caused a 90% decrease in the ACTH-like activity of their samples.

Thus the activity measured by this cytochemical bioassay mimicked that of ACTH in the following ways:

1) It produced a dose-related effect which was parallel to that of standard preparations of ACTH.
2) Its effect could be largely nullified by a specific antibody to ACTH.
3) Its presence in the circulation could be suppressed by cortisol.
4) Its half-time in the circulation was approximately that of ACTH as measured by a different procedure.

IV. Comparison with Other Assays

A standard preparation of ACTH was added to horse serum to give four large samples each containing the hormone at approximately one of the following concentrations: 200, 50, 5.0 and 0.5 μU/ml. The activity of 10 pg of the preparation of the hormone was equivalent to approximately 1 μU. Each of these large samples was assayed at least six times by each of four procedures, namely by the Lipscomb and Nelson (1962) bioassay, which measures the production of corticosterone in vivo; by radio-immunoassays with the antibody directed either to the C-terminus or to the N-terminus; and by the cytochemical segment assay (Rees et al. 1973 b). At the two higher concentrations all assays measured the added hormone and there was no significant difference between the results they recorded. At the concentration of 5.0 μU/ml (or about 50 pg/ml) the bioassay done in vivo could no longer measure the hormone; results by radio-immunoassay and cytochemical bioassay were in agreement. However, at the level of 5 pg/ml, both radio-immunoassays did not consistently measure the ACTH although it was fully assayed by the cytochemical procedure which, of course, was still some way from its limit of measurement (Rees et al., 1973 b; Holdaway et al. 1974 b). Thus in a static situation the cytochemical bioassay measured the concentration of ACTH identically with a bioassay done in vivo and with radio-immunoassay.

V. Comparison with Radio-immunoassay

In contrast to the static situation, which has just been discussed, discrepancies between the cytochemical bioassay and radio-immunoassay were marked in conditions where there was a flux of the hormone. This is to be expected because it is in just such circumstances that the different half-times (or half-lives) of the hormone, recorded by bioassay and by radio-immunoassay, become of critical importance. The typical example in which there is a flux of ACTH is during the insulin hypoglycaemia test which is potentially capable of allowing the assessment of the functional integrity of the entire hypothalamic-pituitary-adrenal axis in man and is used at least as a test of pituitary-adrenal function. Stress, induced by the hypoglycaemia following an injection of insulin, causes the secretion of ACTH, which, in turn, stimulates the secretion of cortisol. Both these can be measured (Fig. 7.7); adequate measurement of the ACTH releasing factor, along the lines of the procedure of Buckingham and Hodges (1977a) would test the whole of the hypothalamic-pituitary-adrenal function.

In one study (Fleisher et al. 1974) the standard preparations of ACTH used for bioassay and for radio-immunoassay were assayed against each other to allow for comparison. They were standardized against the Third International Working Standard, the potency of which is taken as approximately 100 IU/mg (or 10 pg ≡ 1 μU). In samples taken before the onset of stress, the results of radio-immunoassay and of the cytochemical bioassay correlated well (*r*, 0.942) although the former almost always read slightly higher than the latter. Sixty minutes after beginning the test the results still correlated, but radio-immunoassay often gave twice the value recorded by the bioassay. This was the period which showed the greatest discrepancy between the two assays and it is also the period just after the peak concentration of circulating ACTH induced by the hypoglycaemic stress and as measured by radio-immunoassay. If that peak is reached between 45 and 60 min after the test begins (Donald 1971; Fleisher et al. 1974), then this is the period at which some subjects will have passed the maximum in the flux of ACTH and some will just be reaching it. Radio-immunoassay and bioassay will agree as regards the latter; however, if the half-time of bioreactive ACTH is only about 10 min (e. g. Meakin et al. 1959) they will disagree as regards the former because, to the bioassay, perhaps half the ACTH which was present at the peak (at 45 min) will now be inactive whereas much, if not all, the ACTH will still retain its full immunoreactivity. Ninety minutes after the beginning of the test the results of both forms of assay correlated well (*r*, 0.937), with radio-immunoassay generally reading higher than

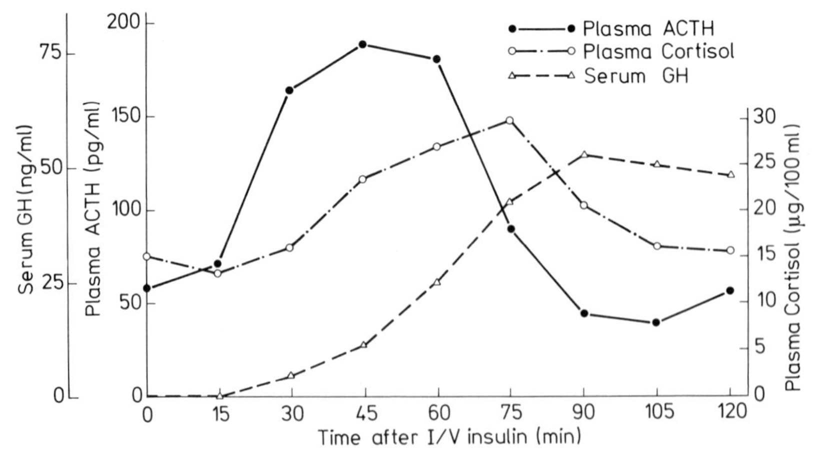

Fig. 7.7. A typical insulin-hypoglycaemia test as measured by radio-immunoassay. (Chayen et al. 1976, p. 77)

bioassay. This again is expected from the different half-times in the circulation of immunoreactive and bioreactive ACTH.

Considerable discrepancies between radio-immunoassay and the cytochemical bioassay (and indeed other forms of bioassay) have been reported in the assay of ACTH produced by tumours which originate from non-endocrine tissues. The biosynthesis of hormone-like molecules by non-endocrine tumours has been reviewed by Rees (1975). Although these molecules behave as immunoreactive ACTH they are often of the big ACTH type (Gewirtz and Yalow 1974) which may have only 3% of the biological activity of α^{1-39} ACTH (Gewirtz et al. 1974) and occur together with biologically inactive ACTH-like fragments (e. g. Bloomfield et al. 1977; Rees 1977). Indeed the considerable discrepancy between the bio- and immunoactivity of elevated levels of ACTH in such circumstances can provide an additional diagnostic indication of the presence of a tumour.

Chapter 8

The Section Assay of Adrenocorticotrophin

A. Background

I. The Need for a More Routine Assay

The cytochemical segment assay of ACTH (Chayen et al. 1971 a, 1972 a) readily found acceptance among European endocrinologists who were concerned with this hormone. Thus it was now possible to measure the biologically active hormone, even when it occurred at concentrations lower than those which normally circulate in the human, without extracting the plasma (as was necessary for the isolated-cell assay of Sayers et al. 1971). The considerable sensitivity of the assay made it ideal for studies done on small laboratory animals and for sequential sampling in the human, as will be discussed in Chap. 9.

For most of these purposes the segment assay was adequate. Its inadequacy was a poor through-put which was not serious as long as its use was restricted to research purposes. But there was a growing awareness that radio-immunoassay occasionally could be inadequate, in that there were times when the concentration of immunoreactive ACTH might be misleading as regards the degree of biological activity actually present. This, after all, was the reason that the WHO committee had called for 'biological microassays' which could be done in parallel with radio-immunoassay (WHO Report 1975). Thus with the ready acceptance of the principle of cytochemical bioassays, and with the increasing worry over possible discrepancies between analytical (radio-immunoassay) and functional assays (as discussed in Chap. 2), there came a demand for a form of cytochemical bioassay which could be used even as part of the routine work of an endocrine laboratory.

The problem in increasing through-put was as follows. Because each assay was a within-animal assay, all the tissue had to be derived from the same animal; even after the 5-h culture period, the cells in the zona reticularis of each guinea-pig responded somewhat differently from those derived from other animals. Consequently a dose-response calibration graph (as in Fig. 3.3) had to be constructed for each guinea-pig; this required four segments. Yet it was rare to obtain more than six manageable segments from the two adrenal glands from any one animal; it was sometimes possible to cut the glands into quarters, making eight segments in all. But with four segments needed for the calibration graph, this left only two (and, rarely, four) with which to assay the samples of plasma. Since it is essential to assay each sample at two dilutions (as discussed above), it was possible to assay only one, or at the most two, samples of plasma on the adrenal glands of one guinea-pig. The next sample required another dose-response calibration graph for the new guinea-pig used for this assay, and so on. Thus, assuming six segments from each guinea-pig, to assay ten samples (each at two dilutions) required ten guinea-pigs, and sixty sets of measurements, that is four segments for each calibration graph plus two for the two dilutions of the plasma, for each sample analysed. As a feat of research it might have been commendable, but it was far too laborious to contribute appreciably to clinical research or practice. It was also unsuitable for the many samples, taken from small laboratory animals, which physiologists and pharmacologists, such as Buckingham and Hodges (1975, 1976), had to assay.

II. The Problems

It is quite obvious that the only way of retaining the benefits of within-animal assays and yet increasing the number of assays which could be done in unit time, was to make the segments very much smaller. In theory it was possible to cut thick slices, about 200 μm thick, and try to make the relevant cells in these respond to the hormone. After all, this is a normal procedure in much research on hormones. But it still would entail sectioning such slices in order that the cytochemical reactions be done and measured. It was therefore logical to try to cut histologically acceptable sections, let us say 20 μm rather than 200 μm thick, and to test the responsiveness of the cells in such sections. Although this was logical all experience was against its possibility. Firstly it required that the chilling and cold-sectioning of the tissue, which had to precede the exposure of the cells to the hormone, would not unduly disturb the cell membranes, and their associated biochemical mechanisms. Everyone knew, or thought they knew, that ‘freezing’ tissue and cutting ‘frozen’ sections, ‘killed’ cells. Secondly we were well aware that fresh sections, reacted at physiological pH values, lost much (or even most) of their contents into the reaction medium (as discussed in Chap. 5). How could one distinguish loss of ascorbate due to such causes from the specific ascorbate-depletion induced by ACTH? Admittedly there were now ways of keeping cells intact during such reactions, but these involved the use of colloid stabilizers which would also stabilize the cell membranes against any perturbation induced by the hormone. All in all the problems seemed virtually insuperable and tribute must be paid to my colleagues, particularly Dr. J. Alaghband-Zadeh, for their pertinacity in the face of so many obstacles.

It was first found that when the cells of the zona reticularis, in sections of the adrenal gland, were exposed to ACTH they did respond, in that the amount of ascorbate lost from the cells increased with increasing concentration of the hormone. The response improved when the thickness of the sections was increased up to about 20 μm. The largest dimension of the cells of the zona reticularis in these cryostat sections did not exceed 18 μm, and the best response was obtained with sections which were 20 μm thick. Protection of such whole cells was not as great a problem as protecting cut cells as would occur, for example, in sections 10 μm thick. Moreover, in previous studies on whole blood-borne cells it had been found that a commercially available polypeptide, derived from the degradation of collagen (Polypeptide 5115), was a good stabilizer of cells, even at relatively low concentrations, without having the strong, apparently over-stabilizing, influence of the various polyvinyl alcohols. This was tried on the sections of the adrenal gland and, at only 5% concentration, produced an immediate improvement in the reproducibility of the dose-response in these sections without impairing the sensitivity of the response. It may be emphasized that to achieve any section assay, a delicate balance must be sought between sufficient stabilization to retain the integrity of the cells (and their biochemical systems) and over-stabilization which stops the ability of the cells to respond to the hormone.

When sections were immersed in the T8 culture medium alone, maximum response was achieved at 30 s, with a symmetrical rate of recovery (Fig. 8.1 a). This was too rapid for routine use, and the maximum was too sharp for good reproducibility in that if some cells were a little slower, or some a little faster, the variance between cells would be large. Addition of ascorbate to the T8 medium slowed the response (Alaghband-Zadeh et al. 1974) and gave it better characteristics (Fig. 8.1 b). The influence of ascorbate in the culture medium indicated, but by no means proved, that some Mass Action effect might be involved.

Finally it was found that if the tissue had been primed with a tenth of the minimum concentration of the hormone which could be detected by the assay, the full sensitivity of the segment assay was regularly obtained even in these sections. The significance of this priming is not known. Certainly it does cause some perturbation of the cell membrane, as shown in the study on its effect on the cell membrane enzyme, 5′-nucleotidase (Chambers and Chayen 1976). It also appears to improve the hydration of the cells (unpublished interferometric observations), which may be important when these are to be chilled and then flash-dried, as they are when sections are produced.

The result of all these manoeuvres was a section assay which was as sensitive and as precise as the segment assay from which it originated (Fig. 8.2). Thus while there is no evidence concerning whether or not the cells were still 'alive' after they had been chilled, sectioned, flash-dried and rehydrated, they did retain their normal structure (Fig. 8.3) and they did respond with full sensitivity to the hormone. At least these findings indicate that the procedures used for preparing sections for cytochemical analysis do little damage to the cells, so validating the use of these procedures for cytochemical studies generally.

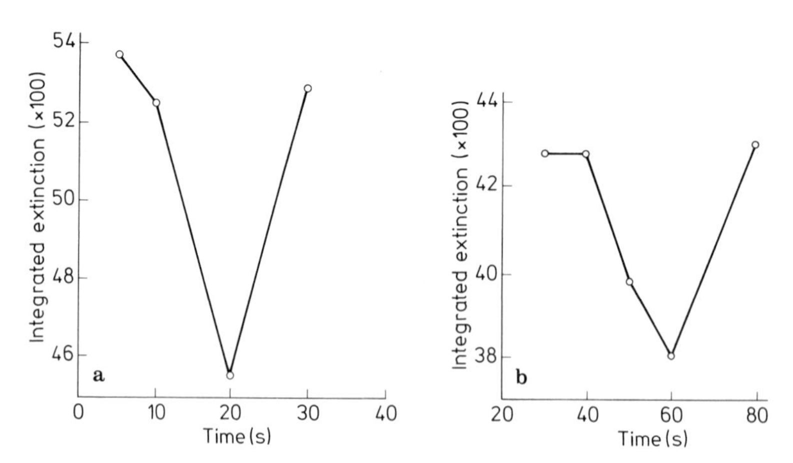

Fig. 8.1 a, b. The effect of ascorbate in the medium in the section assay. **a** The response of sections (20 μm) to ACTH (5 pg/ml) at different times of exposure to the hormone, in the presence of 5% of Polypeptide 5115; **b** The response when ascorbate (10^{-3} *M*) was included in the medium to which the sections were exposed to ACTH (5 pg/ml). (Alaghband-Zadeh et al. 1974, p. 323)

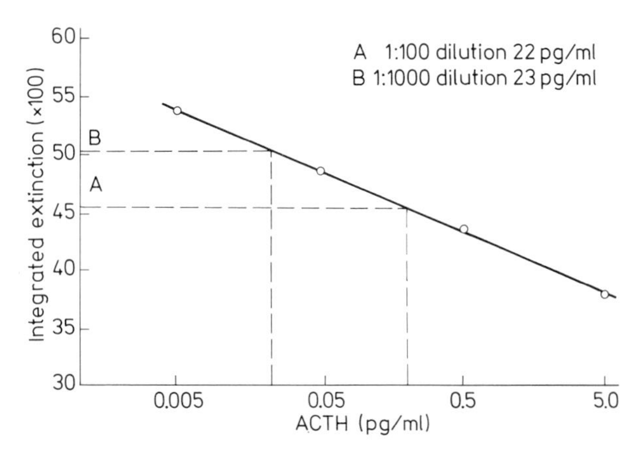

Fig. 8.2. A typical calibration graph and assay done by the section assay for ACTH. The *solid line* shows the dose-response to graded concentrations of the standard preparation. The sample has been assayed at two concentrations; the integrated extinction recorded for these concentrations has been read off the graph (*broken line*). (Alaghband-Zadeh et al. 1974, p. 322)

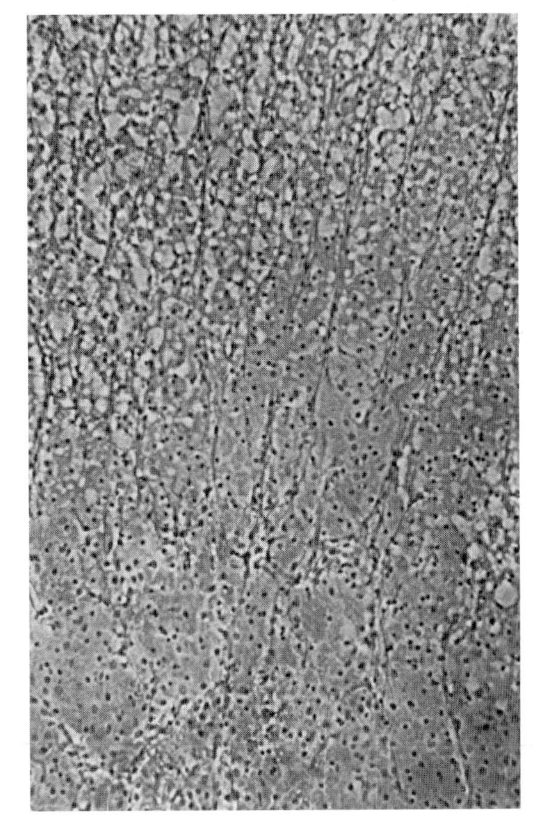

Fig. 8.3. A phase-contrast photomicrograph of a section of guinea-pig adrenal cortex after the section had been exposed to the hormone in the section assay of ACTH. There is no discernible histological distortion. (Chayen et al. 1976, p.66) × 300

B. Cytochemical Section Assay of ACTH

The adrenal glands are removed from one guinea-pig (of about 500 g body wt; the Hartley strain, of either sex, has been used in this work) and each is cut in half. Each half is maintained separately in vitro in the Trowell system (as described in Chap. 4; Appendix 1) with the Trowell's T8 medium supplemented with sodium ascorbate (10^{-3} *M*). It is advisable to maintain each segment separately because the segments secrete cortisol into the medium during the 5-h culture period and this, and any other steroids secreted into the medium, may influence the subsequent response of the cells to ACTH (Chayen et al. 1974 b). After 5 h the medium is replaced with fresh medium containing 0.5 fg/ml of the hormone for 4 min. Then the segments are chilled to $-70°$ C and stored at this temperature. They must be used within 3 days of being chilled. They are then sectioned at 20 μm, the sections are flash-dried in the usual manner and may be kept in a vacuum desiccator (over calcium chloride as the desiccant) for a few hours at room temperature before being used. They are then clipped in pairs, back to back, in the lid of the assay trough (Fig. 8.4). The trough is divided into compartments so that, when the lid is fitted on to the trough, each pair of slides fits into an individual compartment. The first four compartments contain the standard preparation of the hormone dissolved at each of the four concentrations required for the calibration graph (5 fg to 5 pg/ml inclusive). The next two compartments contain the first plasma sample, at 1/100 and 1/1000 dilutions; the next two are used similarly for the second sample, and so on; two can be used for a quality control sample, also tested at two dilutions.

The lid, with the sections clipped to it, is applied to the trough so that the sections are exposed to the hormone for 60 s. The lid is removed and transferred to the cytochemical reaction trough, which contains the freshly prepared ferricyanide-ferric solution (Appendix 3) for demonstrating the presence of reducing moieties, so

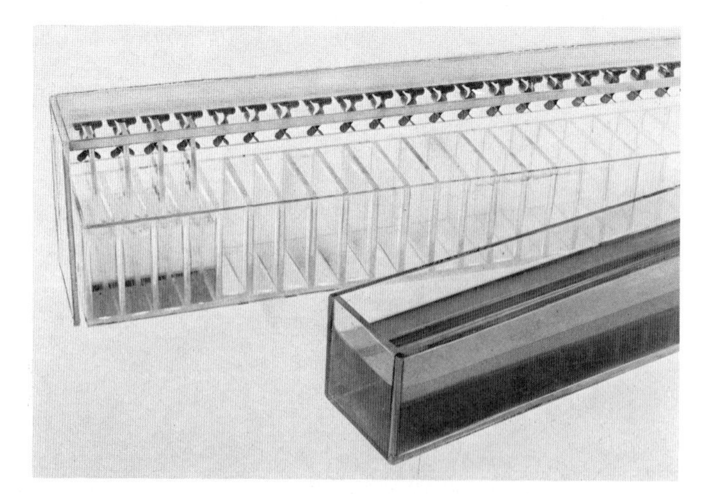

Fig. 8.4. The apparatus for the section assay of ACTH. The sections are held in duplicate, back to back, in the loose-fitting lid so that each pair is located within a separate compartment of the trough. In the foreground is the cytochemical reaction trough, containing the ferricyanide-ferric solution, to which the sections will be transferred by moving the lid from the first to the second trough. (Chayen et al. 1976, p. 64)

plunging all the sections simultaneously into the acidic reaction medium, which also stops the action of the hormone. After 5 min, the lid is transferred to a second reaction trough containing fresh ferricyanide-ferric solution and this process is repeated until the cytochemical reaction has proceeded for a sufficient time (e.g. 15 min). The sections are then rinsed in tap water to stop the reaction. They can then be left, in the dark, to dry, after which they are cleared in xylene and mounted in a suitable mountant, such as DPX or Mercia mount.

The amount of blue colour in the cells of the zona reticularis, produced by the reduction of ferricyanide and the formation of ferric ferrocyanide, is measured at 680 and at 480 nm (as for the segment assay) by means of a Vickers M85 microdensitometer. Normally a $\times 20$ objective can be used, with a mask that encompasses five to ten cells per field; the size of the scanning spot is not critical.

C. Validation

The reproducibility of the procedure was examined by measuring the reaction product in five or six sections exposed to each concentration of the standard preparation of ACTH (5 fg/ml to 5 pg/ml inclusive) and to two concentrations of a plasma. Over this whole range of concentration the coefficient of variance did not exceed 5%. In a series of studies by Buckingham (1974) the slope was 5.2 ± 0.02, index of precision was 0.062 ± 0.002 and the variance was 0.105 ± 0.01 (n, 10). Similar results have been obtained by my own colleagues. A known concentration of a standard preparation of ACTH, added to a plasma which had been assayed for its content of this hormone, gave recovery of 102% and 101%; excess of an antibody against ACTH (Wellcome anti-human ACTH serum, RD 05–06) reduced the ACTH-like activity of plasma by 93% (Alaghband-Zadeh et al. 1974). The assay of plasma samples by the segment and section assay systems gave similar results (Alaghband-Zadeh et al. 1974). This section assay has been used fairly extensively in other laboratories (e.g. Hodges and Vellucci 1975; Buckingham and Hodges 1977a, b, c; Reader et al. 1976; Bloomfield et al. 1977) and seems established as a routine bioassay of ACTH. One worker can produce up to 40 assays each week, which is a considerable advance on the cytochemical segment assay.

Chapter 9

Application of Cytochemical Bioassays of Adrenocorticotrophin

A. Physiological Studies

I. Half-Time of Bioreactive ACTH

The half-time in the circulation of biologically active corticotrophin endogenous to the subject was calculated from a study in which a male subject was given 100 mg of cortisol intravenously and the levels of ACTH were measured at frequent intervals (Rees et al. 1973 b). The results showed that the cortisol suppressed the secretion of ACTH, the levels falling from 43 pg/ml to 34 fg/ml (Fig. 9.1). This gave a half-time, over the first part of the graph, of 10.4 min. In other studies (Daly et al. 1974 a) where only 5 mg of cortisol or 0.5 mg of dexamethasone were used to produce the suppression, the half-times were only about 7 min. It may be remarked that, because the assay is so sensitive, only very small samples of blood need to be taken so that it was perfectly feasible to remove the seven samples required for Fig. 9.1 without incommoding the subject.

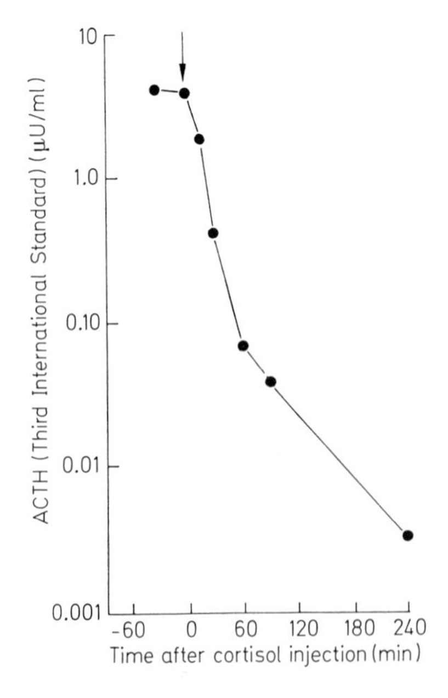

Fig. 9.1. The suppression by cortisol of circulating levels of ACTH-like activity, as measured by the cytochemical bioassay. At the time indicated by the *arrow*, 100 mg of cortisol were injected intravenously into a human subject. The rate of disappearance of endogenous circulating ACTH can be measured. (Rees et al. 1973 b)

The half-time in the circulation of exogenously added hormone was measured in studies on male and female subjects in whom the secretion of ACTH had been suppressed with dexamethasone (2 mg at midnight and 2 mg at 0.800 hours prior to the investigation). Various amounts of the α^{1-24} and of the substituted α^{1-18} ACTH were injected intravenously and samples of the circulating blood were taken, through an indwelling needle in the opposite arm, at 2–4 min intervals for the first 30 min and then at 15 min intervals during the next 2 h. A typical result with the α^{1-24} ACTH is shown in Fig. 3.7. The peak concentration of the 800 ng of the α^{1-24} ACTH was found regularly 4 min after the injection and the concentration assayed at this time gave a fair recovery of the amount injected, taking into consideration the blood volume of the subject (which was only an estimated value). The rise in the concentration of cortisol in the blood followed the peak concentration of ACTH, about 20 min after the injection. The half-time of the injected ACTH, calculated from the steep decline, was about 9 min (Daly et al. 1974 a). In contrast, the long-acting substituted α^{1-18} ACTH (Maier et al. 1971; Keenan et al. 1971) took longer to reach its peak concentration: in one subject injected with 500 ng a plasma level of 195 pg/ml was reached 8 min after the injection; in another, given 120 ng, the peak level of 36 pg/ml was achieved after 12 min, and a peak of 25 pg/ml was recorded in this subject 15 min after injecting 60 ng (Daly et al. 1974 a). The concentration in the blood of this substituted ACTH remained high for a considerable time (Figs. 3.8, 3.9) even when only 60 ng were injected to simulate the normal circulating level of the normal form of corticotrophin. The cortisol levels also remained elevated, in accord with the prolonged retention of this form of corticotrophin in the circulation.

When the substituted α^{1-18} ACTH was tested in the cytochemical bioassay it gave a normal time-response graph, so that it could be assayed in the normal way, using the substituted material for deriving the calibration graph. As has been discussed in Chap. 3, such in vitro bioassays can tell you only the *amount* of the bioreactive material present. They cannot tell you whether their duration of action is altered, probably because of a different rate of degradation in the human body. That is not the function of bioassay; this type of information belongs to physiology and can be obtained very simply by the sort of physiological study, aided by bioassay, which has just been described.

II. Circadian Rhythm and Episodic Secretion of ACTH

The circadian rhythm was measured by assaying for ACTH and for corticosteroids at 08.00 and 20.00 hours in 14 patients in hospital who had no signs of endocrine or neurological disease and were not suffering from a stressful condition. The mean levels of ACTH were 70 (range: 40–100 pg/ml) and 24 (range: 5–45) pg/ml at these times respectively. The corticosteroid levels were 17.7 and 6.8 μg/100 ml (Daly et al. 1974 a). Daly and his colleagues also measured these hormones every 20 min in a sleeping subject who was monitored by electroencephalography. (For procedure, see Daly and Evans 1974). As found by Krieger et al. (1971) and by Gallagher et al. (1972, 1973), they showed that ACTH was secreted in an episodic, or pulsatile, fashion; in this study at least they demonstrated that the peaks of secretion of ACTH preceded those of cortisol; the height of the peaks of ACTH became greater from 3 a. m., reaching a maximal height (75 pg/ml) at about 8 a. m. Further studies on the feedback regulation of the secretion of ACTH (Reader et al. 1976) regarding the effect of cortisol infused at relatively normal levels, indicated that minor increments in the concentration of cortisol in the blood produced rapid modulation in the secretion of ACTH. However, in the absence of exogenous cortisol, no clear pattern was found between the episodes of secretion of ACTH and the level of cortisol in the blood. As Reader et al. (1976) commented, further more sophisticated studies are still required to fully clarify this relationship.

B. Advantages of a Very Sensitive Assay

I. ACTH Levels in Neonates

One of the advantages of a highly sensitive bioassay is that it requires only a small volume of plasma (e. g. 20 μl), seeing that the plasma is to be diluted at least one hundred times before being applied to the target cells in order that the concentration of the hormone shall be in the linear range of the assay. This factor was used by Holdaway et al. (1973) for the assay of ACTH in neonates. Samples were taken (between 9 and 10 a. m) from 14 healthy babies (aged 7–10 days) at the same time as a heel-prick sample was taken for a routine Guthrie test. The mean concentration of ACTH in the blood of these babies was 63 pg/ml (range 13–137); this was significantly higher than the values found in ten normal adults who were accustomed to venepuncture (30 pg/ml, range 8–50, at 9 a. m., falling to 8 pg/ml at 10 a. m., range 1–17 pg/ml). However, it is not clear whether these higher values, found in the neonates, were due to the stress associated with the heel-prick or to a true difference in their hypothalamic-pituitary activity (Holdaway et al. 1973).

II. Hypopituitarism

The second advantage of the sensitivity of these assays is that they can be used to explore the physiology of subnormal levels of the hormone. Holdaway et al. (1973) took blood, under basal conditions, from six patients who had clinical evidence of severe hypopituitarism and in whom the levels of plasma corticosteroids were not detectable by fluorimetric assay. The concentration of ACTH in these samples was 0.35 pg/ml (range 0.017–0.96). The plasma from a patient with a steroid-secreting adrenal tumour assayed at 0.08 pg/ml. In three of the patients with hypopituitarism who were subjected to an insulin hypoglycaemia test (see below) all showed significant increases in ACTH as a response to the stress, the increases being 4.5, 11.2 and 2.2 times greater than the basal levels. Thus in these patients, it was possible to measure low levels of ACTH, which could not be measured by any other assay, and these levels responded to stress, indicating that even at such levels the assay was measuring ACTH and that this assay can be used to investigate hypothalamic-pituitary-adrenal function even in such patients.

III. Studies on Small Animals

Another potential advantage of a sensitive bioassay is that it can be used in physiological and pharmacological studies on small laboratory animals. Whereas previously such investigations required blood samples from several animals for each test, with all the interanimal variation that this entails, it is now possible to use the blood from a single animal; theoretically at least, it should be feasible to take serial samples from one animal subjected to various experimental conditions. Studies of this type have been made mainly by Hodges and his co-workers (e.g. Hodges 1976). They include investigations of the influence of reserpine (Hodges and Velucci 1975) or of the nature of the mechanism by which glucocorticoids, such as betamethasone, control the hypothalamo-pituitary adrenocortical function in the rat (Buckingham and Hodges 1974, 1975, 1976, 1977b).

C. Insulin-Hypoglycaemia Test of Hypothalamic-Pituitary-Adrenal Function

I. Insulin-Hypoglycaemia Test

The measurement of the ability of the hypothalamic-pituitary-adrenal axis to respond to stress, such as surgery, may be of vital importance. It has been of value in the diagnosis of Cushing's syndrome (James et al. 1968) and of hypopituitarism (Landon et al. 1966). Of all the tests, probably the most valuable is that involving stress evoked by insulin-hypoglycaemia (Jacobs and Nabarro 1969).

A common method (e.g. Landon et al. 1963; Fleisher et al. 1974) involves the subject being fasted overnight before an indwelling needle is inserted into a vein in the forearm and the blood sampled. Thirty minutes later 0.15 unit of soluble insulin for each kg of body weight is injected. Samples of blood are taken immediately before injection, and 30, 60 and 90 min later. (It should not be necessary to say that this procedure should be done only under the supervision of a physician and with facilities for injecting glucose should the subject show an unacceptable degree of stress.)

In normal subjects the effects of the hypoglycaemic stress are as follows: First there is a decline in blood glucose. This prompts the stimulation of secretion of ACTH and, shortly thereafter, of growth hormone. Finally the secretion of cortisol is seen to be enhanced (Fig. 7.7). The results of the test are normally assessed by the maximum incremental rise in the level of growth hormone, of cortisol and of ACTH. In one study (Daly et al. 1974b), in 12 patients with rheumatoid arthritis (who had not been treated with steroids or ACTH) the mean level ($\pm$SD) of glucose fell to 21 ($\pm$10) mg/100 ml; the increment in serum growth hormone was 52 ($\pm$29) ng/ml; the incremental rise in plasma ACTH (by the cytochemical bioassay) was 129 ($\pm$63) pg/ml and in plasma corticosteroids it was 16 ($\pm$8) μg/100 ml.

II. Activity-Time Response in the Hypoglycaemia Test

The insulin-hypoglycaemia test had been used to determine whether or not corticosteroid therapy, particularly for patients with rheumatoid arthritis, impaired the hypothalamic-pituitary-adrenal function. It had been shown that the incremental rise in endogenous corticosteroids was unaffected in rheumatoid patients receiving ACTH but was decreased in those given exogenous glucocorticoids. Yet the former treatment should have mimicked the effect of the latter. Thus it was suggested (Daly et al. 1974b) that although ACTH might have suppressed the function of the pituitary (and there was some evidence to this effect in that the incremental rise in growth hormone was impaired), the exogenous ACTH could have caused some degree of adrenal hyperplasia and hypersensitivity which would compensate for what was assumed to be a diminished secretion of ACTH in response to stress. The obstacle to testing this hypothesis (Daly et al. 1974 b) was the difficulty in determining the concentration of ACTH in the blood, especially when this was low. With the advent of the cytochemical bioassay of ACTH, this problem was reinvestigated (Daly et al. 1974 b).

When the maximum incremental rise was used as the criterion for this test, it was clear that in patients treated with ACTH, the incremental rise was greatly diminished although the incremental rise in cortisol was unchanged. (In patients given corticosteroids, both were markedly impaired.) This apparent paradox was resolved when it was realized that *any* appreciable increase in the circulating concentration of ACTH is liable to induce the secretion of cortisol. Thus the significant phenomenon is not the incremental rise but the amount of ACTH secreted and the time over which this level remains capable of acting on the adrenal glands. That is to say, the factor which should be measured is not the maximum concentration which can be achieved, but the area of the whole concentration-time graph (Fig. 9.2). When this was calculated from the results from these patients (Daly et al. 1974 b) it was found that there was no statistically significant difference between the total amount of ACTH secreted by these rheumatoid patients (on ACTH therapy) and that secreted by the control group. Consequently it was not surprising that the total amount of cortisol secreted by these two groups was equivalent, seeing that the time-course for cortisol was much slower than that of ACTH. Thus the apparent paradox was resolved. However, it must be noted that its resolution depended on measuring biologically active ACTH, and the biological half-time of this hormone in the circulation. The area under the curve of

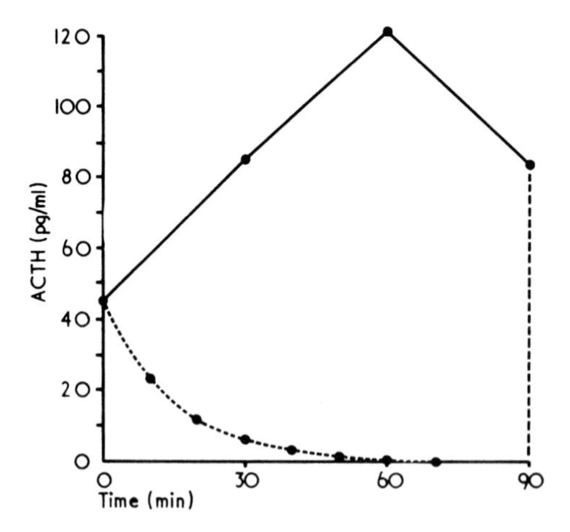

Fig. 9.2. The area of the concentration-time graph from a patient subjected to the insulin-hypoglycaemia test. The total area, bounded by the *solid* and *broken lines*, is measured. Ideally the test should have been extended; the area has been cut-off at 90 min. The dotted line from 0 to 60 min demarcates the area subtracted (based on the half-time of biologically active ACTH) from the total area to correct for the ACTH that had been secreted before insulin was injected. (Daly et al. 1974 b)

immunoreactive ACTH might not have yielded such clear results simply because the half-time of immunoreactive ACTH bears no relationship to the secretion of cortisol.

Another interesting feature came out of this study (Daly et al. 1974 b). In patients treated with ACTH, the incremental rise was smaller than in the controls, but it was sustained for longer. Thus it would be preferable to sample for up to 2 h to obtain a full activity-time relationship. But the interesting feature was that, with a lower incremental rise, the *rate* of secretion of ACTH was significantly slower than in the control subjects. Thus it may be of value to measure the *rate* of secretion in such tests and the cytochemical bioassay, requiring only small samples of blood, is ideal for such measurements. Moreover, the assessment of the whole hypothalamic-pituitary-adrenal function would benefit from measurement of the secretion of the corticotrophin-releasing factor (CRF) as discussed in Chap. 14.

Chapter 10

The Assay of Thyroid Stimulators

A. Background

I. Thyroid Stimulating Hormone

1. Nature of Thyrotrophin

As reviewed by Hall et al. (1975), human TSH is a glycoprotein which, even when relatively pure, may contain five active components; one of these, after purification, had a molecular weight of 34,000 (Roos et al. 1975 a). Like the other glycoprotein hormones it consists of two chains, of almost equal molecular weight, the α-subunit, containing 14% carbohydrate, being very similar in all these hormones. The hormonal specificity seems to reside in the β-subunit which, in the case of human TSH, contains about 6% carbohydrate. Although the isolated subunits appear to have little or no biological activity, they can react in immunoassays; it is remarkable that the α- and β-subunits can be recombined under mildly alkaline conditions to yield material which may have as much as half the original biological activity (Pierce 1971).

Up to the present, efforts to produce highly purified standard reference preparations have not been fully successful apparently because the highly purified hormone is relatively unstable as regards its biological activity, even though it may be suitable for immunoassay (WHO Report 1975, pp. 14–15). For example, the stability of the biological activity of the Medical Research Council (MRC) reference preparation 68/38 hTSH is too poor for it to be used as a standard for biological activity at the concentration at which it normally is tested, but it is an acceptable reference standard for radio-immunoassay. The less pure MRC-A standard of human TSH has been used in all the cytochemical work done by my own colleagues and by other laboratories employing cytochemical bioassays.

2. Effect of TSH Acting on the Thyroid

The many biochemical changes induced in thyroid follicle cells in response to TSH have been reviewed in detail by Dumont (1971). Some of these changes required remarkably high concentrations of the hormone. Only those which have formed a basis for a bioassay will be considered here.

There is little doubt that TSH causes the cells lining the thyroid follicles to take up, or endocytose, colloid. Within the thyroid follicle cells, thyroglobulin is split from the colloid to yield the iodinated thyroid hormones (thyroxine and tri-iodothyronine); the lysosomes of the follicle cells are implicated in this process (Wollman 1969; Dumont 1971). The original bioassay of Adams and Purves (1955), done with guinea-pigs, and its more generally used modification in which mice are used (McKenzie 1958 b), depend on the release of radioactive iodine under the stimulation of TSH. So too do the in vitro bioassays of Kirkham (1962) and of Brown and Munro (1967). Some of the oldest methods depended on measuring the increase in the number of droplets of colloid that was induced by the hormone (as reviewed by Brown 1959; also see Shishiba et al. 1967). However, the limit of sensitivity of even the McKenzie in vivo bioassay is about 0.1 mU/ml (Krieger 1974).

Thyrotrophin stimulates adenylate cyclase activity in the thyroid follicle cells; this phenomenon has been used for assaying thyroid stimulators, particularly the thyroid-stimulating immunoglobulins (e.g. Kendall-Taylor 1972; Onaya et al. 1973; McKenzie and Zakarija 1977).

3. Radio-immunoassay

Although receptor assays have been developed (e.g. Mehdi and Nussey 1975; Smith and Hall 1974; Manley et al. 1974) these have not proved as sensitive as radio-immunoassay (Hall et al. 1975), which at best can achieve a sensitivity of about 0.5 μU/ml. Despite this sensitivity few, if any, radio-immunoassays can consistently distinguish the low levels of TSH found in hyperthyroidism from those present in normal subjects (Hall et al. 1975). The frequency histogram of TSH levels in a defined population (Tunbridge et al. 1977) indicated a skew distribution presumably due to the inability of the radio-immunoassay to measure normal values of less than 0.5 μU/ml. Thus for all its sensitivity, radio-immunoassay appears to be incapable of measuring low, but normal, levels of this hormone (also see Scanlon et al. 1978 a). Thus there was need for a more sensitive bioassay.

The relatively poor sensitivity of radio-immunoassay has stultified many studies on the normal physiological behaviour of TSH. As emphasized by Daly (1977), this insensitivity of radio-immunoassay hinders studies on agents that suppress the concentration of circulating TSH or that cause slight changes in circulating levels that are within the normal range. It has also made it impossible to explore the lower limit of the normal range. Thus there is much TSH physiology which cannot readily be investigated by current radio-immunoassays, although much has been done (see Scanlon et al. 1978 b).

4. Discrepancies Between Radio-immunoassay and Physiology

Krieger (1974) reported a study on a young boy whose physiological condition was compatible with hypopituitarism associated with secretion of biologically inactive forms of ACTH and of TSH. Even though this boy had decreased thyroid function, he had elevated immunoassayable levels of TSH. The thyroid was physiologically functional in that it responded to exogenous TSH. No thyroid antibodies were detected. Illig et al. (1975) reported elevated immunoassayable TSH in many children with hypothalamic hypopituitarism. In six, the basal levels of TSH were high, and an exaggerated increase in immunoassayable TSH was found after the intravenous administration of thyrotrophin-releasing hormone (200 $\mu g/m^2$ body surface area). Thus these physiologically euthyroid patients had excessive amounts of immunoassayable TSH which, it may be assumed, had impaired biological activity. Equally Faglia et al. (1975) commented on several patients with hypothalamic-pituitary disorders who, although physiologically and biochemically euthyroid, showed no increase in circulating tri-iodothyronine after treatment with TRH. These authors suggested that the immunoassayable TSH released by such treatment had low biological activity. (For experimental evidence of reduced biological activity of the TSH in such conditions see Faglia et al. 1979, and as discussed in this chapter Sect. B. III.7.)

Thus from 1975 onwards there has been an increasing awareness that there can be discordance between the circulating levels of TSH, as assayed by radio-immunoassay, and the physiological state of the thyroid. Some workers (as cited above) had suggested that the discrepancy could be due to a form of TSH which was measurable by radio-immunoassay but which had less biological potency than the normal hormone, or even lacked biological activity. But such a suggestion could be no more than mere speculation because it was impossible to assay the biological activity of the immunoassayable TSH, seeing that even the most used bioassay (that of McKenzie) had a lower limit of sensitivity of about 100 μU/ml (Krieger 1974) whereas the normal circulating level of TSH is probably closer to 1 μU/ml (Scanlon et al. 1978 a).

II. Thyroid-Stimulating Immunoglobulins

The occurrence of a long-acting thyroid stimulator (LATS), which was not TSH, was signalled by an abnormal response, caused by sera from patients with Graves' disease, in in vivo bioassays of TSH done in the guinea-pig (Adams and Purves 1956) and in the mouse (McKenzie 1958 a). There is now a vast, and somewhat complex, literature on this subject of immunoglobulins that influence the metabolism of the thyroid gland (e. g. as discussed by McKenzie 1968; McKenzie and Zakarija 1977; Hall et al. 1975). Because the stimulator is an immunoglobulin, or a family of immunoglobulins, it was widely assumed that the response would be species-specific. Consequently the stimulation by LATS-positive sera, observed mainly in the mouse, was a manifestation of the influence of only one representative of this class of immunoglobulins, all of which were capable of stimulating the human thyroid gland to produce thyroid hormones outside the normal control by the pituitary gland. It now seems likely (McKenzie and Zakarija 1977) that there is wide cross-reactivity and that the influence of these thyroid-stimulating immunoglobulins or antibodies (TsIg or TsAb) may be detected with thyroids from species other than the human, provided that the methods used are sufficiently sensitive to detect what may be weak cross-reactivity.

The methods of assaying these thyroid-stimulating immunoglobulins (McKenzie and Zakarija 1977) include the following: the stimulation of adenylate cyclase activity in slices of human thyroid gland; their ability to compete with TSH in a TSH-receptor assay with membranes derived from human thyroid glands; the formation of colloid droplets in slices of human thyroid gland in vitro. They have also been assayed in terms of the 'long-acting thyroid stimulator-protector' (LATS-P) (Munro 1977). Based on comparative results from a variety of such methods, and on the cytochemical bioassay (which will be discussed later), McKenzie and Zakarija (1977) concluded that Graves' disease is due to a circulating polyclonal immunoglobulin (also see Volpé et al. 1974) that is capable of stimulating the human thyroid gland by interaction with an homologous antigen. However, the immunoglobulin can also cross-react with a similar molecule in the thyroid of a distant species. When it cross-reacts with the mouse thyroid so strongly that it causes an appreciable response in the McKenzie mouse assay it is, by definition, *the* long-acting thyroid stimulator, or LATS-positive; if its cross-reactivity with the thyroid of the mouse or of some other species is demonstrable only by other, possibly more sensitive, procedures, it is LATS-negative. Although it is unlikely that this view will be uncontested, it forms a useful hypothesis for the present monograph.

At comparable concentrations, thyroid-stimulating immunoglobulins appear to act more slowly, and to achieve their maximal stimulation later, than does TSH (McKenzie and Zakarija 1977). Thus, for example, Shishiba et al. (1967) found that the appearance of colloid droplets in the thyroid follicle cells occurred later when the thyroid was stimulated by LATS than by TSH; this finding is relevant to the cytochemical bioassay of these stimulatory immunoglobulins.

B. Cytochemical Segment Assay of Thyroid Stimulators

I. Rationale

It is well known that when TSH binds to the base of the thyroid follicle cell it induces endocytotic activity at the opposite surface, so allowing the cell to take up colloid. Lysosomes fuse with the endocytic vacuoles and discharge their hydrolytic enzymes into the phagolysosomes which result from this fusion (Wollman 1969; Dumont 1971), to act on the colloid to produce thyroxine or tri-iodothyronine. The movement of lysosomes towards the lumenal border and the alteration in their surface properties which allows them to fuse with the endocytic vacuole, which is bounded by a pinched-off fragment of the cell membrane, must also be under the control of the stimulus set-up as a consequence of TSH binding to the non-lumenal surface (Fig. 10.1). It was therefore reasonable to investigate the nature of the change in the lysosomal membrane and whether this change was quantitatively related to the concentration of TSH acting on the cell surface. And, as discussed in Chap. 5, cytochemistry is ideally suited for such investigations.

However, there was another indication that lysosomal function could be a valuable marker in the study of thyroid stimulators. It had been shown that an antibody-antigen reaction at the surface of ascites tumour cells (occurring in the absence of complement and hence with no cytotoxicity) produced a marked but transient labilization of the membranes of lysosomes deep in the cytoplasm (Dumonde et al. 1961, 1965; Bitensky 1963 b). Thus it seemed likely that the effect of binding to the surface of the thyroid follicle cell by thyroid-stimulating immunoglobulins might also influence the state of the lysosomes inside the cytoplasm.

II. Procedure

The procedure (Bitensky et al., 1974) is as follows. Young female guinea-pigs, about 300 g, are used for this assay because the thyroid glands of older animals tend to develop cystic change. The animal is killed by asphyxiation with nitrogen and the two lobes of the thyroid gland are removed expeditiously and trimmed of excessive fat or connective tissue. Each of the two lobes is cut into three segments (Fig. 10.2) and each of these segments is maintained separately in vitro in Trowell's maintenance culture for 5 h at 37° C (for the reasons discussed in Chap. 4). The

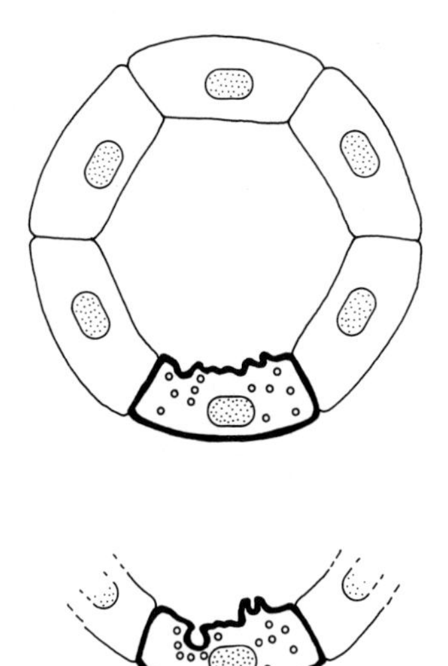

Fig. 10.1. Representation of how the thyroid follicle cell responds to TSH. *Top*, TSH binds to the base of the cell and stimulates endocytotic ripples in the cell membrane on the opposite surface, in contact with colloid. *Middle*, endocytotic vesicles form by the invagination of the plasma membrane and lysosomes move towards the vesicles. *Bottom*, the plasma membrane is pinched-off to yield intracytoplasmic vesicles with which the lysosomes fuse

medium is the unmodified Trowell's T8 medium at pH 7.6; the gas phase is 95% 0_2 : 5% CO_2. The time from the death of the animal to having the segments in culture should not exceed 15 min. At the end of the 5-h period the medium is replaced by fresh T8 medium containing either one of four graded concentrations of a standard preparation of the hormone (e. g 10^{-4} to 10^{-1} μU/ml) or one of two dilutions of the plasma in T8 medium (e. g. at 1 : 100 or 1 : 1000 concentration). After 7-min exposure to the hormone or plasma (for the assay of TSH) the segments are chilled to $-70°$ C. Within a week, the chilled tissue is sectioned at 10 μm in a cryostat at $-25°$ C, with the haft of the knife cooled with solid carbon dioxide (as discussed in Chap. 5) and the sections are reacted for 7 min for lysosomal naphthylamidase activity (see Chap. 5 and Appendix 3; also Bitensky et al. 1974 a; Chayen et al. 1973 b; Bitensky and Chayen 1977). As discussed in Chap 6, because the coloured reaction product occurs mainly in small granules of the size of lysosomes,

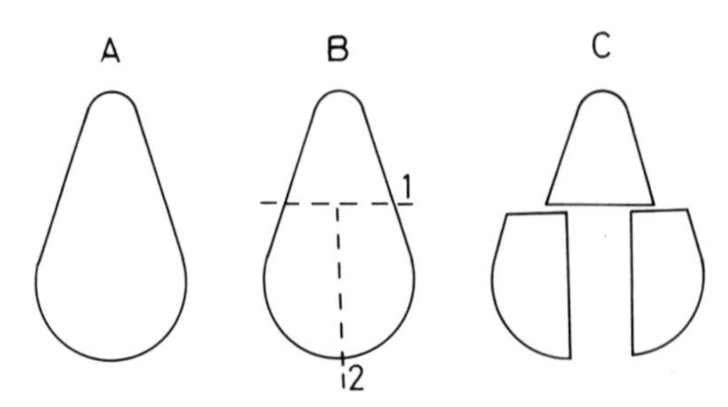

Fig. 10.2. How to cut three segments from each lobe of the guinea-pig thyroid gland

microdensitometry of the enzyme activity in individual follicle cells requires the use of an oil-immersion objective (×100, N.A. 1.2) and a scanning spot of 0.2 μm diameter in the plane of the section; the optimal wavelength is 550 nm. Generally ten readings (equivalent to ten follicle cells), each preferably from a different follicle, are made in each of two duplicate sections. With such small pieces of thyroid tissue, it is best to cut three serial sections for each slide and to measure the activity only in the middle section, which should be free from compression and other deleterious effects of sectioning.

For the demonstration of TsIg activity the graded concentrations of TSH are replaced by dilutions of a LATS preparation which had a known potency in the McKenzie assay. Thus for a sample of which 1 mg produced a 500% response, at 24 h, in the McKenzie assay, dilutions of 0.001 to 1.0 μg/ml were used (Bitensky et al. 1974 a). The time course is also altered to 20 min (Bitensky et al. 1974 a) or, preferably, 30 min (Petersen et al. 1975) in accordance with the slower rate at which these immunoglobulins achieve their maximal response.

III. Validation

1. Time Course of Response

Over the concentrations of TSH and of TsIg used in these studies, the maximal effect of TSH on the labilization of lysosomal membranes in the follicle cells occurs rapidly, with almost a plateau in the time-response graph between 5 and 10 min (Fig. 10.3). In contrast, LATS (Bitensky et al. 1974 a; or the 7S fraction of the serum from a patient with untreated Graves' disease: Petersen et al. 1975) shows a delayed effect, producing little or no activation of the lysosomes after 7-min exposure (Fig. 10.4), but reaching maximal activity about 30 min after the application of the immunoglobulin to the segments. Thyroid stimulating immunoglobulins which are LATS-negative can also be assayed in this way (McKenzie and Zakarija 1977).

2. Nature of the Response

Typically the log dose-response graph is curvilinear (Fig. 10.5) and indicates that TSH increases the manifest, or freely available, lysosomal naphthylamidase activ-

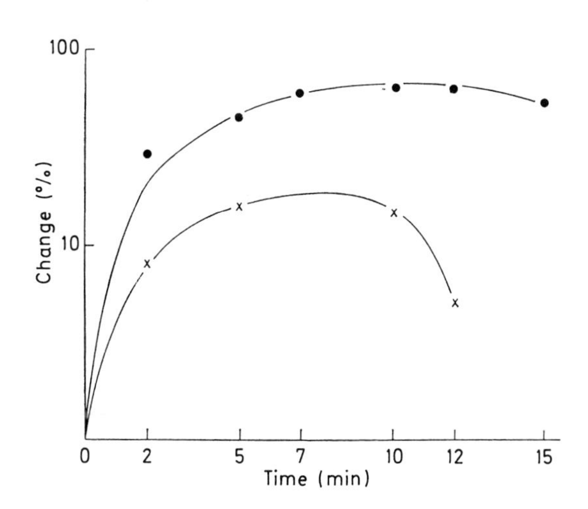

Fig. 10.3. The time course of the response in freely available naphthylamidase activity in segments of thyroid gland exposed to TSH at 4×10^{-2} (*solid circles*) and 4×10^{-4} (*crosses*) μU/ml. (Bitensky et al. 1974 a, p. 366)

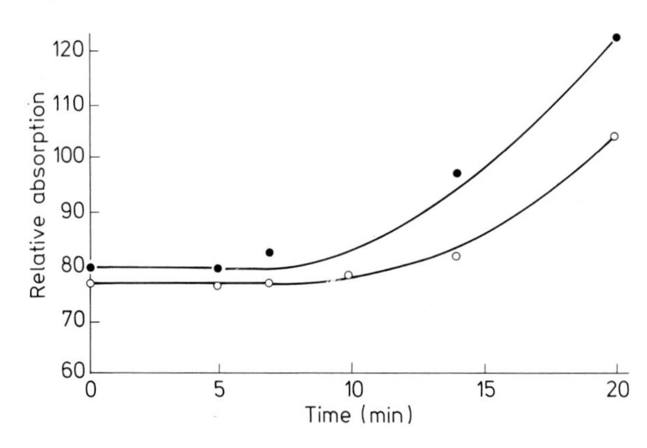

Fig. 10.4. The free naphthylamidase response in segments of thyroid gland exposed to LATS (*crosses*, 1 μg/ml; *circles*, 0.1 μg/ml) for various periods of time. (Bitensky et al. 1974a, p. 371)

ity. Very little of the enzyme activity demonstrated by this cytochemical procedure is due to other 'aminopeptidases' as shown by the fact that inclusion of a preferred non-chromogenic substrate for aminopeptidases, namely leucineamide (Bitensky and Chayen 1977), in the reaction medium causes not more than 10% inhibition of this reaction (Ealey 1979). A further indication that this activity may be lysosomal (see Chap. 5) is that the activity (measured in serial sections) does not increase linearly with time of reaction but shows two different rates, being relatively slow up to 10 min and faster thereafter (Fig. 10.6). This is in accord with the view that there is

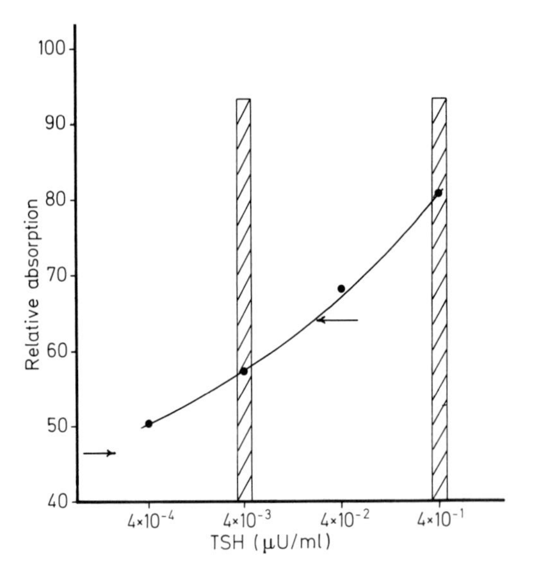

Fig. 10.5. A typical curvilinear dose-response calibration graph for a segment assay of TSH. The *solid circles* show the mean relative absorption due to free lysosomal naphthylamidase activity in follicle cells in segments exposed to the various concentrations of a standard preparation of the hormone. The *hatched columns* show the total lysosomal naphthylamidase activity in sections of two of these segments, demonstrating that the hormone does not change the total activity of the enzyme but only the rate at which the substrate can reach the enzyme. The *upper arrow* indicates the activity induced by a sample of plasma diluted 1 : 100 (the second dilution, at 1 : 1000, is not shown), giving a final concentration (after correcting for dilution) of about 1 μU/ml. The *lower arrow* shows the activity remaining in the same plasma after it had been treated with an antibody specific to the hormone. (Chayen 1978, p. 338)

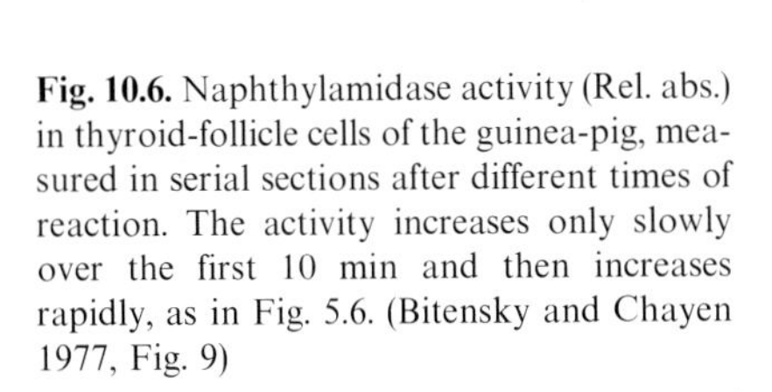
Fig. 10.6. Naphthylamidase activity (Rel. abs.) in thyroid-follicle cells of the guinea-pig, measured in serial sections after different times of reaction. The activity increases only slowly over the first 10 min and then increases rapidly, as in Fig. 5.6. (Bitensky and Chayen 1977, Fig. 9)

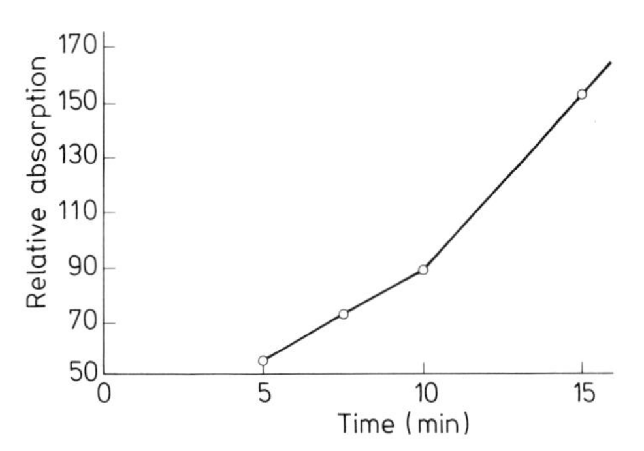

a change in latency of the enzymatic activity during the course of the relatively acidic reaction. It would therefore be self-negating to use a reaction time of, let us say, 15 min if we were concerned to measure changes in lysosomal enzyme latency induced by the hormone or immunoglobulin. The reaction time normally used is 7 min, which is sufficiently long to produce a measurable reaction product even in relatively unstimulated cells, but which is still just within the slow rate of reaction, the implication being that the lysosomal membranes are still exerting a restraint on the entry of substrate to the lysosomal enzyme (as discussed in Chap. 5).

Proof that the changes in activity recorded in the assay are due to changes in the permeability of the lysosomal membranes is the fact that pretreatment at 37° C of serial sections, prior to the naphthylamidase chromogenic reaction, in an acetate buffer at pH 5.0 for various times, causes an increase in activity. In relatively unstimulated cells the acidic pretreatment, which has been used by de Duve (1963) to disclose total lysosomal activity biochemically, caused little change over the first 6 min but then increased the measurable activity to almost double that which was found without acidic pretreatment (Fig. 10.7). Thus in these lysosomes the latent

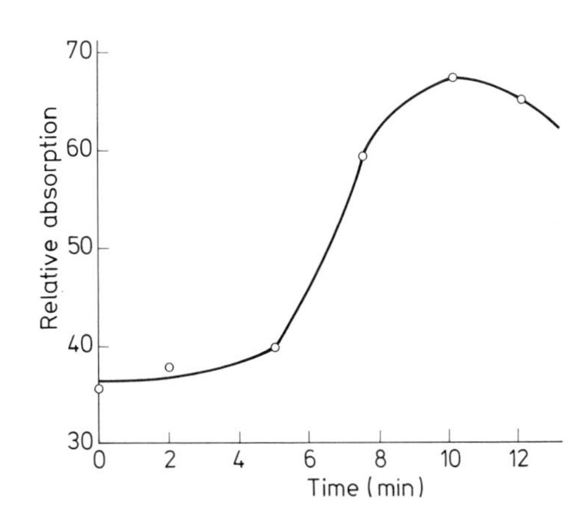

Fig. 10.7. The effect of pretreatment in acetate buffer, pH 5.0, at 37° C, on the activity of lysosomal naphthylamidase in the thyroid-follicle cells of the guninea-pig. In serial sections this enzymatic activity (Rel. abs.) increased with prolonged acidic preincubation until the activity, at 10 min, was almost double the 'freely manifest' activity, disclosed without prior acidic pretreatment. The time allowed for the cytochemical reaction is the same for all these sections. This figure corresponds to Fig. 5.7. (Bitensky and Chayen 1977, Fig. 10)

activity, i.e. 'total' (activity at 10 min) minus 'manifest' (at time 0), was about as much as the freely available activity (measured at time 0). Or, to put the matter differently, the freely available (or manifest) activity in these cells was about 50% of the total activity. By a similar process it can be seen (Fig. 10.5) that the effect of stimulation by TSH is merely to increase the proportion of freely available activity; it did not alter the total activity. Consequently it may be claimed that it has influenced the permeability of the lysosomal membrane, not the enzyme within the lysosomes.

Another point which may bear stressing is also shown in Fig. 10.7. After prolonged activation – in this case by acidic pretreatment – the enzymic activity declines. This is taken to be caused by the lysosomal membranes becoming so modified that the naphthylamidase leaches out into the acidic acetate buffer and hence is no longer available for reaction when the sections are transferred into the chromogenic reaction medium for the demonstration of the lysosomal naphthylamidase activity. Similarly if the concentration of TSH acting on the tissue is excessive, and the lysosomal membranes are fully labilized, some loss of enzyme can occur even in the chromogenic reaction medium and this enzyme, now in solution, will not be available for producing a coloured reaction product inside the section. Hence one limitation of these techniques is that you cannot measure a higher concentration of TSH, or of TsIg, than will produce 100% labilization. This phenomenon can lead to some confusion to the unwary, as in the following situation. Suppose that the graded concentrations of the standard preparation of TSH produce a calibration graph rising from 60 to 100 units of relative absorption, the highest value being the activity stimulated by TSH at a concentration of 10^{-1} μU/ml. Suppose then that the unwary experimenter has ignored the request, made repeatedly in this monograph, that the plasma sample should be assayed at two dilutions, and has used only a 1 : 1000 dilution. Let us say that the naphthylamidase activity recorded by the tissue exposed to this dilution is 90 units of relative absorption (per unit field, per unit time of reaction). It is possible that this value can be read off the calibration graph and measured as μU of TSH activity; it will be a little less than 10^{-1} μU/ml. But it is equally possible that the sample, as applied to the tissue, contained 1 μU/ml and that it over-labilized the lysosomes, causing the enzyme to leak into the chromogenic reaction medium. So, although it fully labilized the lysosomal membranes, allowing the naphthylamidase to show total activity (i.e. 100 units of relative absorption), 10% has also been lost into the medium and therefore was not recorded. Had the experimenter used a 1 : 100 dilution as well, he would have produced a reverse response, the 1 : 100 dilution causing more enzyme to leak out and therefore giving a lower measurement than the 1 : 1000 dilution. This would have alerted the experimenter to the possibility that his plasma sample was insufficiently diluted to fit to the standard calibration graph. (Other causes for such non-parallel results will be considered later.)

3. Specificity and Accuracy

Bitensky et al. (1974 a) and Döhler et al. (1977, 1978) showed that the TSH-induced increase in lysosomal naphthylamidase activity was almost completely nullified when a specific antibody to hTSH was added to the plasma. Bitensky et al. (1974 a) also showed that a preparation of luteinizing hormone (code number 68/40), at a concentration of 5 mU/ml, gave activity corresponding to 7×10^{-4} μU/ml of TSH;

this activity was commensurate with the known levels of TSH present as impurity in this preparation of luteinizing hormone and it was abolished by the antibody to hTSH. So although both are glycopeptides, sharing the same α-subunit, there was no cross-reactivity. Similar results were found with chorionic gonadotrophin (Petersen et al. 1975). Bitensky (personal communication) has shown that the individual α- and β-subunits have very little effect on this assay, or on the section assay. Petersen et al. (1975) demonstrated that the values recorded by this assay were increased in four volunteers injected with TRH (200 μg, iv), the results being in reasonable agreement with radio-immunoassay. Similar results were reported by Döhler et al. (1978). In two subjects given 120 μg of tri-iodothyronine orally each day, the circulating level of the activity recorded by this assay (Petersen et al. 1975) dropped from 0.33 to 0.17 μU/ml on the 4th day, and to 0.018 μU/ml on the 8th day in the first subject; from 0.63 to 0.015 (on the 8th day) in the second subject. It is noteworthy that even such low values were measurable by this asay.

In assessing the accuracy of the procedure Ealey (1978) recovered 88%–95% of the TSH added exogenously (5 – 200 μU/ml) to samples of plasma.

4. Sensitivity and Reproducibility

The lowest concentration on the normal calibration graph that produces a response significantly greater than that of fresh medium alone is about 10^{-4} μU/ml (Bitensky et al. 1974a), 5×10^{-4} μU/ml (Schäfer and Böcker 1976) or 5×10^{-5} μU/ml (Petersen et al. 1975). The lowest concentration that has been measured is 0.015 μU/ml in a T_3-suppressed individual (Petersen et al. 1975). Döhler et al. (1978) measured basal circulating levels of between 0.11 and 0.34 μU/ml in patients with hyperthyroidism (Graves' disease) and these values were annulled by the addition of an antibody specific for hTSH. In these patients, injection of TRH had no influence on the circulating level of TSH, although it had an effect on such patients after therapy, increasing the circulating levels by a factor of five or ten to around 1 μU/ml which was close to the limit of detection of the radio-immunoassay used by these workers.

Petersen et al. (1975) examined the coefficient of variation of the assay at the four concentrations used in their calibration graphs, namely at 5×10^{-5} to 5×10^{-2} μU/ml. In fourteen successive assays they obtained coefficients of interassay variation of 5.8%, 6.6%, 6.8% and 6.5% for these concentrations. Similarly Ealey (1978, 1979) found the intra-assay variance to be $6.5 \pm 3.3\%$; the fiducial limits based on repeated assays of eight samples were 75.3% to 132.8%. She overcame the problem of determining parallelism of two curvilinear responses (namely that of dilutions of the standard preparation and that of dilutions of the plasma) by estimating the maximal acceptable variance between the results obtained with two dilutions of the plasma. A variance of up to 16% was found to be compatible with a parallel response. Responses of the naphthylamidase activity to different dilutions of plasma that, on this basis, were not parallel to the calibration graph were found in samples from patients with various disorders of the hypothalamic-pituitary-thyroid axis. Such cases of non-parallelism deserve to be studied in much greater detail as has been done in connection with the assay of gastrin-like activity (see Chap. 12).

Ealey (1978, 1979) made an assessment of the half-time in the circulation of bioreactive TSH by studying the rate of fall in the circulating levels in patients subjected to the TRH test. In relatively normal patients the biological half-time was

21–25 min. She also showed that although the reference preparation was stable for 140 days if stored suitably at −70° C, the TSH activity in plasma deteriorated by approximately 1% per day when stored at this temperature.

5. Inadvisability of Using Serum

Ealey (1978, 1979) compared the results of bioassaying TSH in samples of plasma and serum taken at the same time. The results with serum were frequently, but by no means invariably, unaccountably anomalous, compared both with the cytochemical bioassay done on the plasma from that patient, and with the radio-immunoassay. Thus it is important to use plasma for the cytochemical bioassay. One probable reason for the anomalous results obtained when serum was used is the fact that polyamines such as spermidine, which can simulate the effect of TSH (see below), occur in blood-borne cells but not in plasma (Documenta Geigy, 1970); the destruction of these cells, to a greater or lesser degree, during the clotting process and the separation of serum can release these polyamines and so influence the cytochemical bioassay.

6. Correlation with Radio-immunoassay

In assays on 18 normal subjects Petersen et al. (1975) obtained close correlation between the results of radio-immunoassay and cytochemical bioassay (the correlation coefficient, r, being 0.95). Ealey (1979) also obtained good agreement by these procedures applied to measuring circulating levels in normal subjects.

It may be noted that big TSH isolated from human pituitary glands was assayed both by radio-immunoassay and by the cytochemical bioassay (Erhardt and Hashimoto 1977) although the biological activity was less than the immunoactivity.

7. Circulating Levels of TSH and Discrepancies with Radio-immunoassay

Although there is reasonable correlation between the results obtained by radio-immunoassay and by the cytochemical bioassay in normal subjects, the fact that the latter is more sensitive means that it can assay the circulating level of TSH in normal subjects even when this is less than the limit of sensitivity of the particular radio-immunoassay used, be that 1.0 or 0.5 μU/ml. Thus Petersen et al. (1975) found that the circulating level of TSH, measured by the cytochemical bioassay, in normal subjects ranged from 0.3 to 3.0 μU/ml. Patients with hyperthyroidism (Graves' disease) had about 0.1 μU/ml which was neutralized when an antibody against TSH was added to the plasma (Döhler et al. 1978). According to von zur Mühlen et al. (1977), 10%–15% of clinically euthyroid patients with goitre show no rise in circulating levels of TSH after injection of TRH. They investigated ten such patients in whom both Graves' disease and autonomous adenoma had been excluded. These had low levels of TSH, as measured by the cytochemical bioassay (0.11–0.34 μU/ml, normal values being about 0.8 μU/ml) and none showed a rise 30 min after TRH had been injected. These results implied an autonomous behaviour of the thyroid gland, or possibly a more efficient gland.

Petersen et al. (1978; also Belchetz and Elkeles 1976), using both radio-immunoassay and the cytochemical bioassay, showed that in patients with hypothalamic-pituitary disease the basal levels of TSH were elevated as measured by the former method, but normal as assayed by the latter. Although these patients had no circulating thyroid antibodies and had normal uptake of thyroidal iodine in response to exogenous TSH, they showed no rise in thyroid hormones in response to TRH. Yet they showed a typical delayed, but marked, TSH response to TRH when the TSH was measured by radio-immunoassay which was very discrepant from the results with the cytochemical bioassay. Thus these workers were able to claim clearly, for the first time, that in these cases there was secretion of TSH which had impaired biological activity, so confirming the opinions of previous workers such as Faglia et al. (1975), Patel and Burger (1973), Krieger (1974) and Illig et al. (1975), and as discussed in Sect. 10.A.I.4.

Marked dissociation of the results by immunoassay and by the cytochemical bioassay has been reported by Faglia et al. (1979) in five patients with central hypothyrroidism due to idiopathic disease, which indicates that, in these cases of secondary hypothyroidism, the immunoreactive TSH had lower biological activity than normal TSH both in basal conditions and after stimulation with exogenous TRH. Thus the cytochemical bioassay of TSH is clarifying the occasional discrepancies that have been found between the high values of circulating TSH, found by radio-immunoassay, and the physiological euthyroid or hypothyroid condition of the subject investigated (Chap. 10, Sect. A.I.4.). In fact, one of the cases of hypothalamic hypopituitarism with high levels of circulating immunoreactive TSH, investigated by Illig et al. (1975), and discussed in Sect. 10.A.I.4., was investigated by the cytochemical bioassay and showed very low biological activity of the high immunoassayable TSH (Bitensky, personal communication).

IV. Studies with Thyroid-Stimulating Immunoglobulins

The completely disparate times of response of the thyroid-follicle cells to usual concentrations of TSH and of the thyroid-stimulating immunoglobulins (Bitensky et al. 1974a; Petersen et al. 1975; also Figs. 10.3, 10.4) allow the measurement of each type of stimulator even in the presence of the other. However, most work has been concerned with the activity of a specific fraction obtained from serum or other fluids. Thus McKenzie and Zakarija (1977) found that both LATS-positive and LATS-negative sera from patients with Graves' disease gave positive responses both in the cytochemical bioassay and in their cAMP assay. McLachlan et al. (1977) used the assay to measure the activity of thyroid-stimulating immunoglobulins present in the immunoglobulin fraction of culture medium in which they had cultured lymphocytes from patients with Graves' disease; the equivalent fraction from lymphocytes from normal subjects showed negligible thyroid stimulating capacity.

C. Cytochemical Section Assay of Thyroid Stimulators

I. Background

The cytochemical segment assay needs six segments of the thyroid gland from one guinea-pig for each assay, involving a four-point calibration graph and two dilutions of the plasma. This means that each segment must be one-third of a lobe of the gland. It was impossible to increase the number of segment assays done on one animal because it is not feasible to work with pieces smaller than one-third of a lobe. As discussed with regard to the assay of corticotrophin (Chap. 8), the only obvious way of increasing the number of assays was to use sections instead of segments. And indeed there was ample need for increasing the through-put of this assay beyond the three, or possibly five, segment assays that could conveniently be done in a normal week. First, the segment assay was far too slow to be used for the second-stage of a two-stage assay of TRH, as will be discussed later (Chap. 14). Second, interesting discrepancies between immunoreactive and bioreactive TSH are being found when the TRH test is applied to certain types of patient. In these tests, where the levels of TSH are monitored following an injection of TRH, the results of one test would require at least 1 week of work by the segment assay. Finally, the cytochemical bioassay may well be required to explore the levels of TSH in that part of the normal population in which the circulating concentrations are below the limits of measurement by radio-immunoassay.

In discussing the development of a section assay for corticotrophin (Chap. 8), some of the problems of protecting sections so that the cells still responded to the hormone have been mentioned. The difficulties of developing a section assay for thyroid stimulators were even greater in that not only was it necessary to protect the content of the cells from becoming dispersed into the medium at a pH of about 7.6, as was required for maximal hormone activity, but it was also essential to protect the lysosomes within the cells. While it was perfectly possible to protect the cells and their organelles (as discussed in Chap. 5), such protection would also protect them from the influence of the hormone. So, for example, in the first studies the lysosomes in sections exposed to the T8 medium became permeable so quickly that the influence of TSH in the medium could not be detected; in sections stabilized as for the section assay for ACTH the lysosomes retained their latency irrespective of the concentration of the hormone in the medium. However, it was known that, at more acidic pH values at least, sections could be immersed in acetate for up to 7 min before the lysosomes became labilized (e. g. Fig. 10.7); moreover, acetate ions even at pH values of about 7 were known to protect protoplasm (e. g. Chayen and Denby 1960). Lysosomes in sections exposed to the hormone dissolved in T8-medium-containing acetate ions did become labilized, but they were also labilized to varying degrees by the acetate medium without the hormone; the medium did not stabilize the sections sufficiently. To achieve slightly more stabilization gum tragacanth was added at very low concentrations; this is another colloid stabilizer, used for example in Turkish delight, and is particularly useful when very slight changes in degrees of stabilization are called for. Finally a concentration was found which gave sufficient stability, in the presence of acetate ions, while not blocking the influence of the hormone (Fig. 10.8). The thickness of the sections was also critical and it may be worthwhile for each laboratory to check the optimal thickness under the conditions prevailing in that laboratory. This is because although the constant drive mechanism on the cryostats (discussed in Chap. 5) ensures constancy of thickness, each cutting speed will produce its own thickness. Consequently even though both you and I may set the advance mechanism to 12 μm, if your automatic drive is moving slower or faster than mine, our sections will be consistently of different actual thickness.

II. Procedure (Gilbert et al. 1977 a)

A small female guinea-pig, of about 300 g, that has been fed with a normal diet supplemented with cabbage daily for at least 8 days, is killed by asphyxiation

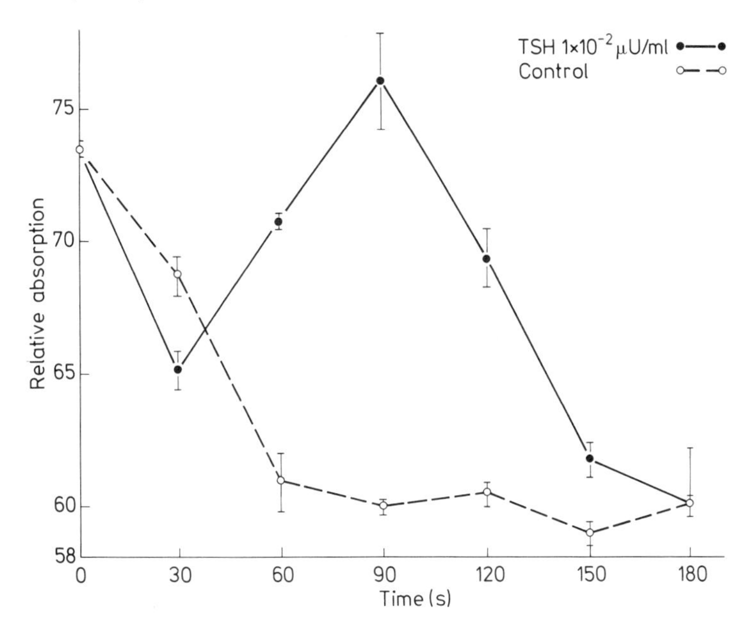

Fig. 10.8. Time course of the response of lysosomal naphthylamidase activity (Rel. abs) in the follicle cells in sections of guinea-pig thyroid gland exposed to TSH (*solid line*) (1×10^{-2} μU/ml) or to the vehicle alone (*broken line*). The activity in both is less after the sections have been exposed for 30 s. The activity in response to the vehicle continues to decline, reaching a stable, minimum value at 60 s. In contrast, the activity induced in response to the hormone increases to reach a maximum at 90 s and declines thereafter. Bars indicate mean values in each duplicate section. The results for each duplicate are marked for each measured point

with nitrogen and the thyroid gland is expeditiously removed and trimmed of excessive fat and connective tissue. Each lobe is bisected and each half-lobe is maintained separately at 37° C for 5 h in Trowell maintenance culture with the T8 medium at a pH of 7.6. As in all these assays, the gas phase is 95% O_2 : 5% CO_2. The tissue is then chilled to −70° C and stored at this temperature.

The tissue remains fully responsive for only 2 days. During this period it is sectioned at 12 μm as has been discussed previously (Chap. 5). The glass slides on which the sections are mounted are clipped in duplicate, back to back, in a lid of a trough very similar to that used for the section assay for corticotrophin. For such assays, to reduce the excessive use of costly T8 medium, it is preferable to have the compartments of small volume and accordingly they are usually made by cutting into a solid base and having a guide on the lid to ensure that the sections locate into the relevant compartment (Fig. 10.9).

The medium contains Trowell's T8 medium at pH 7.6 (adjusted to this pH by adding the appropriate amount of acid to remove bicarbonate and carbon dioxide from the medium) containing 0.05 *M* sodium acetate and 0.02% gum tragacanth. The last is best dissolved in water separately, overnight, at higher concentration (0.1%) and 20 ml of this is added to 80 ml of the T8 medium. (Note: the solution of gum tragacanth should not be stored in a refrigerator.) This complex medium is then used to dilute the standard reference preparation of the hormone or the plasma.

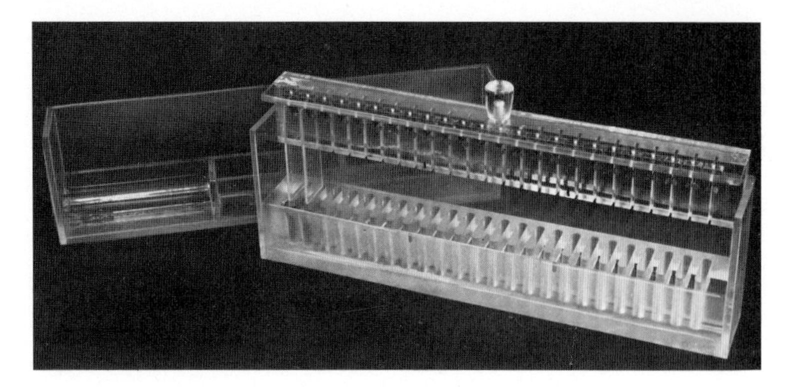

Fig. 10.9. Apparatus for the section assay of TSH

With such small sections it is frequently helpful if three sections are cut, as a ribbon, and affixed to one slide; the middle section should show no deformity and only this section is measured. Two slides, each with three sections prepared in this way, are used, back to back, in each compartment of the trough: the first four compartments contain the standard preparation of the hormone at dilutions of 10^{-4} – 10^{-1} μU/ml; the others contain the various samples of plasma diluted normally $1:10^2$ and $1:10^3$. After they have been dried at 37° C for 10 min, the sections are immersed for 90 s and then the whole lid, with all the sections attached, is moved to the cytochemical reaction trough, so simultaneously immersing the sections in the reaction medium for the demonstration of lysosomal naphthylamidase activity (as for the segment assay; the method of Chayen et al. 1973 b and in Appendix 3). In our hands the optimal time for the cytochemical reaction to give sufficient colour in these thick sections for measurement and yet keep the reaction time to the correct point on the activity-time course (Fig. 10.6) is 6 min.

As was found in relation to the segment assay, concentrations of thyroid-stimulating immunoglobulins generally used in these studies act more slowly and therefore can be assayed by their response at 3.5 min (Fig. 10.10); their effect at 90 s is minimal (Gilbert et al. 1977 a).

As was discussed with regard to the segment assay, the reaction product consists of fine particles and it must therefore be measured with the fullest resolution of which the microscope is capable. Hence it is imperative that a ×100, 1.2 NA objective is used for measuring this reaction product; failure to do so can nullify the assay (Bitensky et al. 1973).

III. Validation

Although the dose-response still has curvilinear tendencies, it is generally sufficiently close to linear to be treated as a linear response (Fig. 10.11). In ten assays, von zur Mühlen et al. (1978) found the response to be linear over the range of concentrations of 0.1 – 0.0001 μU/ml with correlation coefficients of between 0.96 and 0.99; their interassay variation (three assays at two dilutions) was 5.1%. In an as yet unpublished, work by my colleagues (Dr. Lucille Bitensky, Dr. W. R. Robertson and Mrs. D. M. Gilbert) the slope of these assays of TSH is about 7.5, giving indices

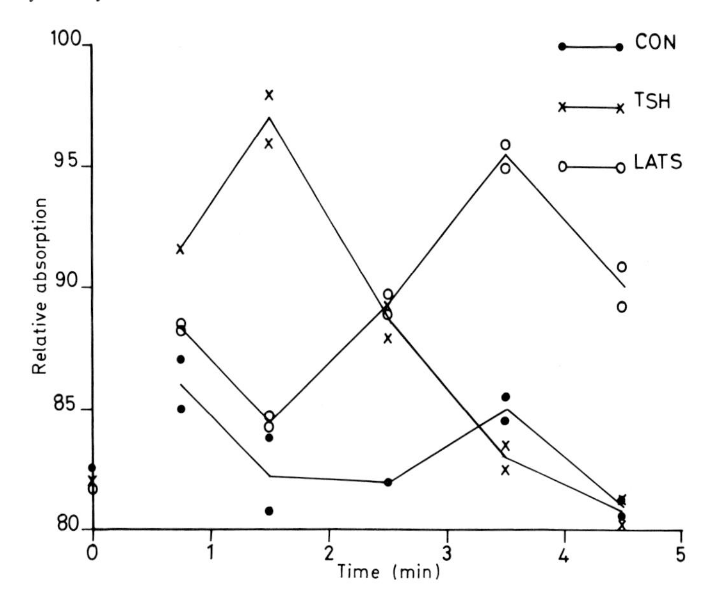

Fig. 10.10. Time course for the freely available lysosomal naphthylamidase activity stimulated by TSH (*crosses*), by a sample containing LATS (*open circles*) and by the vehicle alone (*filled circles*) in follicle cells in sections of guinea-pig thyroid gland. The activity (Rel. abs.) stimulated by TSH reaches a maximum value at 1.5 min and is the same as that caused by the vehicle alone at 3.5 min., at which the response to LATS, at these concentrations, is maximal

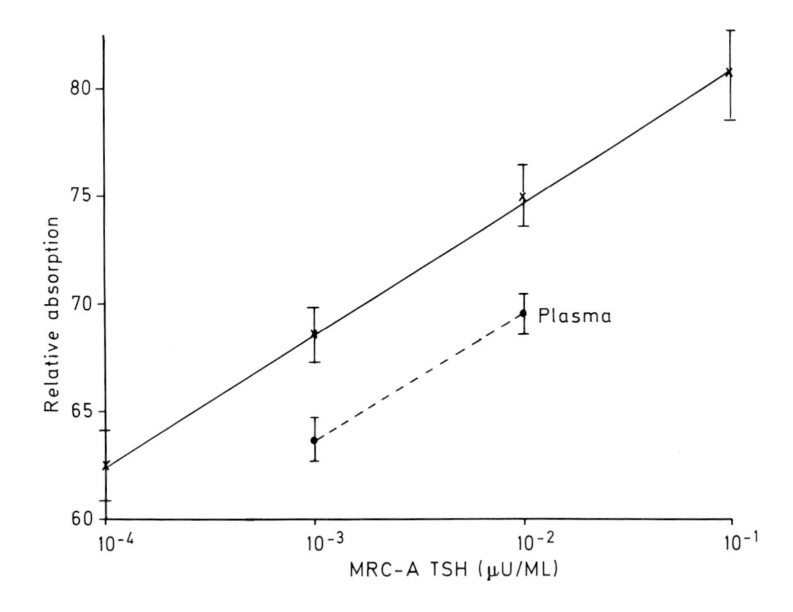

Fig. 10.11. A section bioassay of TSH. The responses (Rel. abs.) to four concentrations of the standard preparation (MRC-A standard preparation of TSH) fall on a linear regression line (*crosses*). The responses to two concentrations ($1:10^2$ and $1:10^3$) of the plasma (*filled circles*) are parallel to those induced by the standard preparation. The concentration of TSH in these two samples is read from the calibration graph. Bars denote mean values in each of the duplicate sections

of precision of about 0.1 and fiducial limits of between 71%–141% and 57%–175%. Full recovery (108%) of TSH added to measured samples of plasma has been obtained. A specific antibody to hTSH, added to a plasma, nullified the TSH activity of the sample (99% inhibition). As was found with the segment assay, luteinizing hormone had only a very slight influence in the section assay, consistent with the known contamination of this preparation by TSH. In accord with this view, the activity was nullified by the addition of an antibody specific for hTSH. Follicle stimulating hormone did not cross-react. Of the two subunits, only the α-subunit had any effect, and then only at very high concentrations due presumably to low levels of contamination by the intact hormone.

The sensitivity of the method, namely the concentration of the hormone that gives an unequivocally greater activity than the vehicle alone, was 10^{-4} μU/ml (Gilbert et al. 1977 a). As regards through-put, our apparatus has 24 chambers, allowing for 4 for the 4 dilutions of the reference preparation from which the calibration graph is prepared, and 20 chambers for 2 dilutions of 9 different samples and of one quality control preparation, if required. But if the material used for quality control is a sample of plasma, it must be remembered that this will lose activity, even if stored at −70° C, by 1% each day of storage (Ealey 1978).

D. A Possible Mechanism for the Action of Thyroid Stimulators

The fact that ACTH very rapidly caused a loss of ascorbate might not be thought surprising. It could be supposed that the very act of attachment of the hormone to the receptor at the surface of the cells of the zona reticularis might cause sufficient perturbation of the cell membrane to allow this loss of apparently soluble material from the interior of the cell. This may be a gross over-simplification, but it is a tenable hypothesis at present; it saves one seeking mysterious mechanisms. But the effect of TSH cannot be dismissed so lightly. When this hormone is applied to segments of thyroid gland it causes a marked change in lysosomal permeability after an exposure of only 2 min (e. g. Fig. 10.3); its effect is maximal 90 s after it reacts with thyroid follicle cells in sections (Figs. 10.8, 10.10). In these times, depending on whether we are discussing the segment or the section assay, the hormone presumably attaches to the receptor at the cell surface; perturbs the membrane with possibly far-reaching consequences; stimulates adenylate cyclase activity and probably other enzymes; and then influences the functional state of the membranes of lysosomes deep inside the cells. The time course seems far too short to involve phagocytosis and the formation of secondary lysosomes, which were discussed in Chap. 10., Sect. B.I. The response seems more akin to that recorded by Dumonde et al. (1961, 1965) when an antibody-antigen reaction, occurring at the surface of ascites tumour cells, produced a transient labilization of the lysosomal membranes inside these cells. The question was: what sort of mechanism could produce such changes, and so quickly?

A clue to a possible mechanism came from a study on the influence of spermidine on lysosomes in thyroid follicle cells in vitro (Gilbert et al. 1977 b). It is well known that polyamines, particularly spermidine, can stabilize the secondary structure of nucleic acids and possibly polyribosomes (Tabor and Tabor 1976); they may also stabilize some cellular and subcellular membranes. Consequently it was hoped that if spermidine were added to the culture medium it might help stabilize lysosomal membranes. Instead it was found that, when added to the medium for the full 5 h, it caused a log dose-related labilization of the lysosomal membranes at concentrations of 100 *pM* to 1 μ*M*.

This clearly called for more detailed investigation. It is known (Richman et al. 1975) that TSH stimulates thyroid ornithine decarboxylase; that this enzyme has a very short half-life (about 10 min in regenerating liver: Tabor and Tabor 1976); and that it is very rapidly activated (e. g. Russell et al. 1970; Byus and Russell 1975; Tabor and Tabor 1976). Thus it was likely that, in addition to activating adenylate cyclase, the hormone could also be stimulating ornithine decarboxylase. This enzyme is the first of a series of enzymes involved in the synthesis of polyamines (Fig. 10.12). The immediate product of its action is the formation of putrescine. The direct effect of this polyamine on the lysosomes of thyroid follicle cells was found to be very slight (Bitensky et al. 1978 a, b). In contrast, spermidine produced a marked effect, in both segments and sections. In the latter system, whereas TSH (10^{-2} μU/ml) produced its maximal response at 90 s, spermidine (5×10^{-7} *M*) gave equivalent labilization at 60 s (Fig. 10.13).

The synthesis of spermidine from putrescine (Fig. 10.12) requires the production of *S*-adenosylmethylthiopropylamine from *S*-adenosylmethionine (SAM) and this process is mediated by SAM decarboxylase. A very effective inhibitor of this enzyme

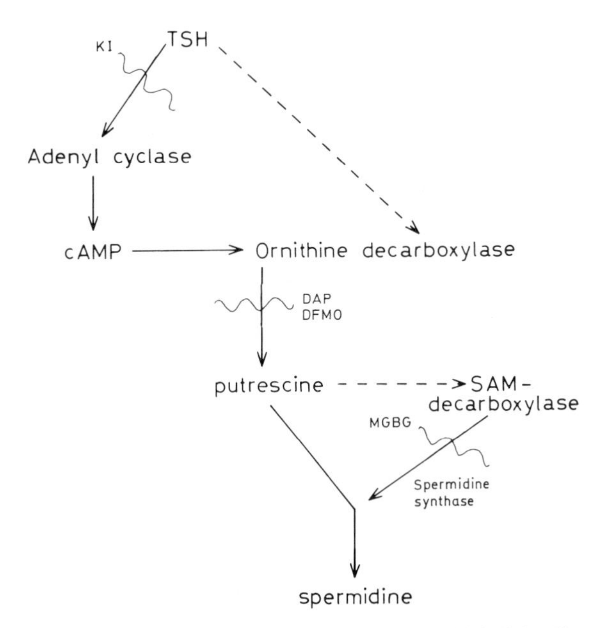

Fig. 10.12. The pathway by which TSH may stimulate the synthesis of spermidine in thyroid follicle cells. TSH stimulates adenyl cyclase activity. (This can be blocked by potassium iodide, KI.) It also stimulates ornithine decarboxylase activity, either through the action of cAMP or perhaps directly. Ornithine decarboxylase activity generates putrescine (which can stimulate SAM decarboxylase). This conversion can be blocked by inhibitors of ornithine decarboxylase (DAP or DFMO). The conversion of putrescine to spermidine requires the activity of both SAM decarboxylase and spermidine synthase. Inhibition of the former by MGBG stops the production of spermidine. The addition of any of the inhibitors shown in this scheme inhibits the increased lysosomal naphthylamidase activity normally induced by TSH. Points of inhibition are shown by a *wavy line*. Details of the inhibitors are given in Chayen et al. (1979)

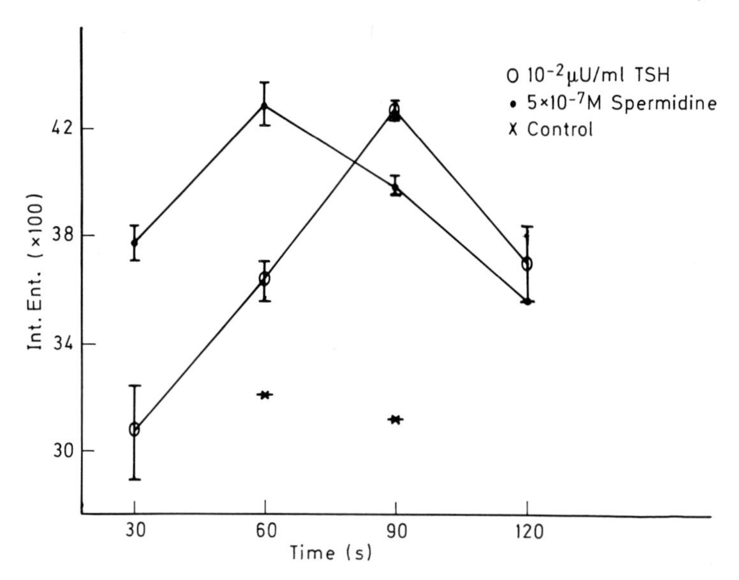

Fig. 10.13. Time course for the activation of 'freely measurable' lysosomal naphthylamidase activity (I. E. × 100) in the follicle cells of serial sections of the thyroid gland of a guinea-pig under the influence of spermidine (5×10^{-7} *M*: *filled circles*), TSH (10^{-2} μU/ml: *open circles*) and the vehicle alone (*crosses*). Maximal activity is induced by spermidine at 60 s; it takes 90 s to achieve maximal activity with TSH under these conditions. Bars denote mean value in each duplicate section

is methylglyoxal bis-(guanyl hydrazone) dihydrochloride (MGBG), which has been used in many studies, both in vitro and in vivo (e. g. Corti et al. 1974; Sturman 1976). In the presence of this inhibitor, even if TSH stimulated the activity of ornithine decarboxylase to produce putrescine, no spermidine would be formed, and putrescine only slightly influences the functional state of the lysosomes in the thyroid follicle cells. Bitensky et al. (1978 a, b) found that when segments of thyroid were maintained for 5 h in the presence of this inhibitor (at 10^{-5}, 10^{-6} and 10^{-7} *M*) and were then exposed to TSH (10^{-1} μU/ml), the response of the lysosomal membranes was inhibited (by 68%, 50% and 10% respectively); complete inhibition of the response was found with the unduly high concentration of MGBG of 10^{-3} *M*. This was not a direct effect of the inhibitor on the lysosomal membranes because even after the tissue had been maintained in vitro in the presence of the inhibitor, spermidine caused the full normal labilization of the lysosomal membranes. Moreover, the inhibitor blocked the effect of dibutyryl cAMP, which alone mimicked the influence of TSH (Chayen et al. 1979)

The fact that spermidine, at quite low concentrations, can produce the same labilization of the lysosomal membranes as does TSH, but more rapidly, and the evidence that inhibition of synthesis of spermidine blocks the effects of TSH and of cAMP, seem to indicate that the long-range effect of stimulation of the cell membrane could be due to a polyamine such as spermidine. The hypothesis would be as follows: the thyroid stimulator reacts with a specific receptor at the cell surface and causes a perturbation of the membrane. This perturbation is reflected by the activation of various enzymes within, or associated with, the cell membrane, notably adenylate cyclase and ornithine and SAM decarboxylase. There is evidence that cAMP can activate ornithine decarboxylase (e.g. Byus and Russell 1975a; Bacharach 1975). Whether the one stimulates the other in this instance, or whether both events are related to the initial perturbation of the cell membrane, is not clear at present. But it seems certain, both from these indirect findings and by direct examination (Richman et al. 1975), that TSH stimulates the synthesis of polyamines so that a consequence of the binding of thyroid stimulators to the cell membrane can be the intracellular production of spermidine, which labilizes the membranes of the lysosomes and so prepares these organelles for the endocytosis of colloid and the subsequent processing of thyroglobulin to produce the thyroid hormones. It also seems likely that the transient labilization of the lysosomal membranes inside ascites tumour cells which responded to an antibody-antigen reaction at the cell surface, as has been discussed, might also have been mediated by a similar process.

This influence of spermidine on lysosomal function has been discussed at some length because it is a newly discovered phenomenon. However, it should be emphasized that the better known effect of polyamines is in relation to cell growth, probably in stimulating protein synthesis but also in cell proliferation (e.g. Byus and Russell 1975a,b; Tabor and Tabor 1976; Mamont et al. 1976, 1978). Consequently it may be expected that the synthesis of polyamines will be stimulated as part of the trophic effect of trophic hormones. And indeed many trophic hormones acting on their specific target tissues have been shown to stimulate ornithine decarboxylase, which is said to be the rate-limiting enzyme of polyamine biosynthesis (Bacharach 1975). This stimulation has been demonstrated in the ovary under the influence of luteinizing hormone (e.g. Nureddin 1977) and in rat liver stimulated by growth hormone (Jänne et al. 1968).

E. Influence of Thyrotrophin on Oxidative Metabolism

I. Significance of the Pentose Shunt

In many ways the process of endocytosis, for example of colloid, is similar to that of phagocytosis. In such actively phagocytosing cells as macrophages, the act of engulfing matter is associated with a marked increase in pentose-shunt oxidation (e.g. Karnovsky 1962) although the precise function of this oxidation in these cells does not seem to have been ascertained. This then, was one reason for interest in the pentose shunt, or hexose monophosphate pathway, in the thyroid follicle cells under the influence of thyroid stimulators. (Also see Hashizume et al. 1975, on the relation of the pentose shunt to endocytosis of colloid.) Possibly a more obvious reason was the trophic influence of thyrotrophin. The pentose shunt (Fig. 10.14) 'shunts' glucose-6-phosphate away from its normally major pathway, namely the Embden-Meyerhof pathway of normal glycolysis, into the hexose monophosphate pathway. In this system two oxidizing enzymes, glucose-6-phosphate dehydrogenase and 6-phosphogluconate dehydrogenase, convert it first to ribulose phosphate. This can then be altered to ribose phosphate as required for the biosynthesis of the nucleic acids (and therefore also of proteins) or it can be converted by a series of reactions to glyceraldehyde-3-phosphate or to fructose-6-phosphate, which are normal components of glycolysis. Thus this pentose shunt contributes to cell growth by providing ribose and deoxyribose. But its contribution is more far-reaching. The

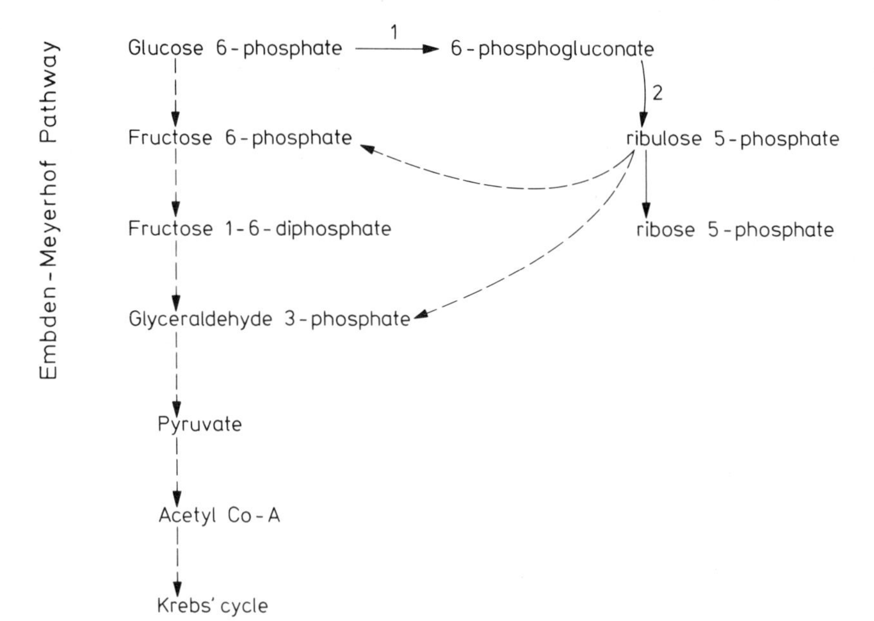

Fig. 10.14. The utilization of glucose-6-phosphate through the Embden-Meyerhof pathway or through the pentose shunt (hexose monophosphate shunt)

two dehydrogenase enzymes require $NADP^+$ as their coenzyme and in oxidizing their substrate they reduce the coenzyme:

$$\text{Glucose-6-phosphate} + NADP^+ \rightarrow \text{6-phosphogluconate} + NADPH.$$

The NADPH so generated can then be used in at least two ways. It can be oxidized by the microsomal respiratory pathway of which cytochrome P-450 is the terminal member. As discussed in relation to the ACTH-induced depletion of ascorbate from the adrenal cortex (Chap. 7, Sect. A.II and III), this pathway is now known not to be exclusive to the microsomes or smooth endoplasmic reticulum and, in general, it appears to correspond to the cytochemists' NADPH-diaphorase. The NADPH used by this system can contribute to hydroxylation, or to other mixed-function oxidations, either for detoxication processes or for the production of specialized molecules. The other, and more relevant, use of NADPH is in the biosynthesis of fats, steroids and proteins (e. g. as discussed by Chayen et al. 1973 a); increased generation of NADPH for this function precedes the synthesis of DNA in proliferating cells (Coulton 1977) and, during fracture healing, accompanies both cellular proliferation and the earliest indication of the formation of bone (Dunham et al. 1977). Thus it may be expected that trophic hormones will stimulate the hexose monophosphate pathway. The enzyme which regulates this pathway is glucose-6-phosphate dehydrogenase (Gumaa et al. 1971; Eggleston and Krebs 1974; Krebs and Eggleston 1974); by its nature it is very susceptible to modification by hormones (as reviewed by Chayen et al. 1974a). The activity of this enzyme can also be modified by less direct factors. It is inhibited by the reduced form of the coenzyme (Eggleston and Krebs 1974) so that the rate at which the NADPH is reoxidized by the diaphorase system could regulate the activity of the dehydrogenase enzyme itself. But the thyroid follicle cells have yet another regulatory mechanism involving NAD^+-kinase, as will be discussed below.

II. NAD^+-kinase

When sections were cut of an unstimulated thyroid gland of the guinea-pig they produced very strong reactions for glucose-6-phosphate dehydrogenase activity provided that $NADP^+$ was included, as it should be, in the cytochemical reaction medium; no activity was present if NAD^+ replaced $NADP^+$. The optimal concentration of $NADP^+$ was 0.28 mg/ml and lower concentrations gave less activity (Fig. 10.15). However, in sections from stimulated glands, NAD^+ did allow the demonstration of a highly active glucose-6-phosphate dehydrogenase, indicating the existence of an active kinase which could convert NAD^+ to $NADP^+$ (Macha et al. 1975, following Dumont 1971). Thus here we have another way of regulating pentose-shunt activity by modulating the activity of glucose-6-phosphate dehydrogenase (Fig. 10.16). In further studies (Bitensky et al. 1976; also unpublished results of Dr. W. R. Robertson) it was shown that this kinase activity was dependent on both ATP and on magnesium and that, given sufficient time for all the reactions to occur, ADP and phosphate could replace ATP. Thus sections given glucose-6-phosphate and NAD^+ showed very little reaction whereas serial sections given

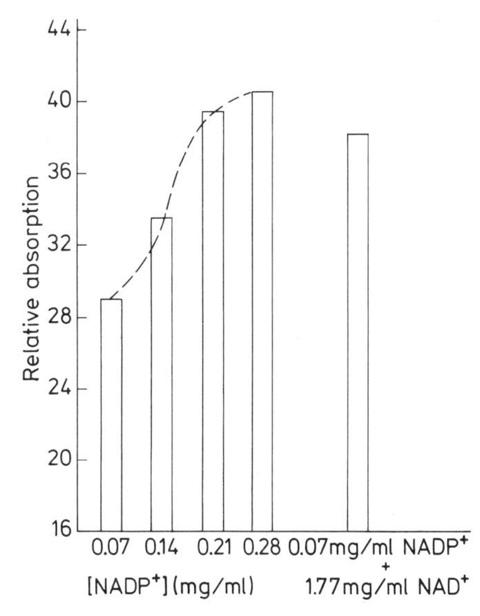

Fig. 10.15. Glucose-6-phosphate dehydrogenase activity in thyroid follicle cells activated either by increasing concentrations of exogenous $NADP^+$ or through the NAD^+-kinase activity on NAD^+. The activity in the presence of 1.77 mg/ml of NAD^+ (together with 0.07 mg/ml of $NADP^+$) was comparable to what would have been obtained with $NADP^+$ at a concentration of 0.2 mg/ml; this is taken to be due to the conversion of NAD^+ to $NADP^+$ by the kinase. (Macha et al. 1975, p. 333)

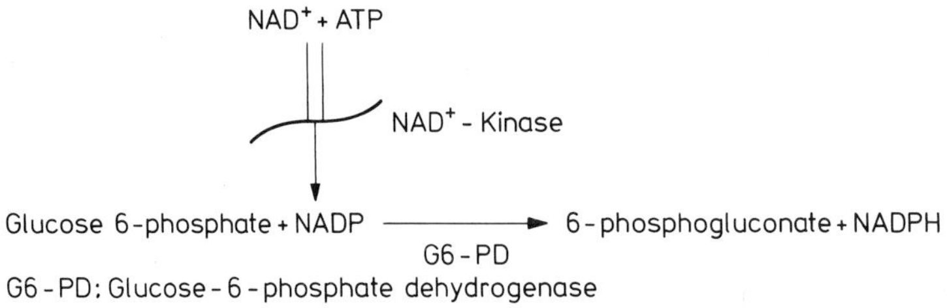

NAD^+ + ATP

NAD^+ - Kinase

Glucose 6-phosphate + NADP ⟶ 6-phosphogluconate + NADPH
G6-PD
G6-PD: Glucose-6-phosphate dehydrogenase

Fig. 10.16. NAD^+-kinase as a mechanism for controlling glucose-6-phosphate dehydrogenase activity. Although the latter enzyme may be in an active form, its activity is curtailed by the amount of $NADP^+$ available to it. The concentration of $NADP^+$ can be increased by NAD^+-kinase phosphorylating NAD^+ to form $NADP^+$

NAD^+ and ATP and magnesium produced strong 'glucose-6-phosphate dehydrogenase' reactions; they produced similarly strong reactions, after an initial lag period, if given ADP and phosphate and magnesium. It seems probable, therefore, that these results, which confirm those of Dumont (1971), demonstrate the presence of a very active TSH-dependent NAD^+-kinase in thyroid follicle cells, and also a system for generating ATP which can be stimulated by TSH.

III. Influence of Thyrotrophin

1. NADPH diaphorase

Macha et al. (1975; also Bitensky et al. 1976) maintained segments of guinea-pig thyroid gland in vitro for 5 h and then exposed some of these segments to various concentrations of TSH (MRC A reference preparation) for different periods of time. They found that TSH at 10^{-2} μU/ml caused little increase in NADPH-diaphorase activity over that in the untreated segments until the hormone had acted for about 20 min, after which they found a marked increase in this activity, reaching about twice the initial activity at 28 min exposure. Lower concentrations (e. g. 10^{-4} μU/ml) produced less stimulation but the maximum was attained earlier (e. g. at 20 min). These results therefore confirmed those obtained by Dumont (1971) who required rather higher concentrations of the hormone to establish this effect, namely that TSH increases the rate of reoxidation of NADPH and, in this way, can increase the actual activity of glucose-6-phosphate dehydrogenase (and hence of the whole pentose shunt) in the follicle cells by removing inhibitory NADPH and making more $NADP^+$ available to the enzyme.

2. NAD^+-kinase

In segments of thyroid tissue treated in the same way (above), Macha et al. (1975; also Bitensky et al. 1976) showed a powerful effect of TSH on the NAD^+-kinase system. After prolonged exposure to the hormone the tissue was sectioned and tested against glucose-6-phosphate and NAD^+ (in some studies a very low concentration of $NADP^+$ was also included in the reaction medium to maintain a low basal level of activity) and the dehydrogenation was measured, as usual, by the reduction of neotetrazolium. (In these studies menadione was used as the intermediate hydrogen acceptor in place of phenazine methosulphate.) The stimulation of this activity, which was dose-dependent, was marked after the

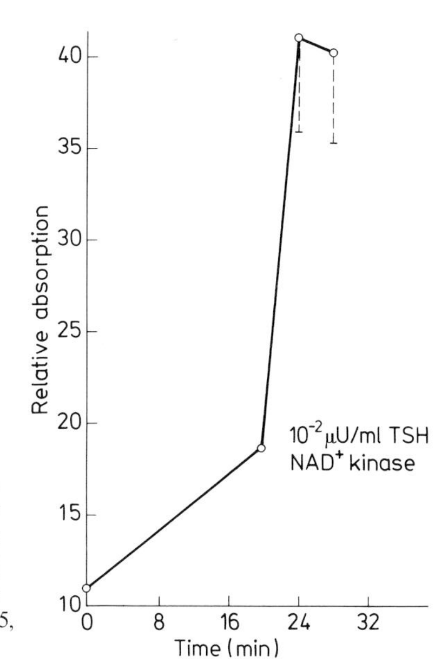

Fig. 10.17. The alteration in NAD^+-kinase activity, measured by glucose-6-phosphate dehydrogenase activity with NAD^+ as coenzyme, in the follicle cells of segments of the same thyroid gland exposed to TSH (0.01 μU/ml) for various periods of time. The *broken lines* indicate the standard deviation of the readings. (Macha et al. 1975, p. 329)

tissue had been exposed to TSH for 20 min, and became maximal at about 24 min (Fig. 10.17). As shown in Fig. 10.15, the activity in such sections indicated a local concentration of $NADP^+$, produced by conversion from NAD^+ by the kinase, of nearly 0.2 mg/ml. These results also confirm the findings of Dumont and his co-workers (Dumont 1971), who required high concentrations of the hormone (128 mU/ml) to demonstrate this stimulation in slices of dog thyroid.

Thus it seems likely that TSH increases pentose-shunt activity by stimulating the rate-limiting enzyme, glucose-6-phosphate dehydrogenase. The particular interest in this mechanism is that apparently these cells have a high concentration of glucose-6-phosphate dehydrogenase in a fully active state but unable to act because it lacks its coenzyme, or because much of the NADP present is in the reduced form because of the slow rate at which it is being reoxidized (by the diaphorase system). TSH releases the enzyme from both restrictions: it increases the rate of oxidation of NADPH and generates ample supplies of the oxidized form ($NADP^+$) by stimulating NAD^+-kinase to produce it from NAD^+. The hormone also may increase the cellular concentration of ATP which will facilitate this conversion. It must be assumed that there are powerful mechanisms for destroying or otherwise inactivating $NADP^+$ in these cells so that when they are not exposed to TSH, as at the end of the 5-h culture period, there is virtually no $NADP^+$ available to the enzyme. If this hypothesis is correct, it makes a fascinating biochemical control mechanism which may be contrasted with the postulated function of ascorbate, in the cells of the zona reticularis, for controlling the activity of the same enzyme, albeit through the respiratory pathway involving cytochrome P-450 (as discussed in Chap. 7, Sect. A. III).

Chapter 11

Luteinizing Hormone

A. Background Information

I. Nature of the Hormone

Human luteinizing hormone (LH), or human lutropin as it is now called, has a molecular weight of about 32,000 (e.g. Roos et al. 1975 b); approximately 28% (w/w) of the molecule is made up of carbohydrate. (Other workers have found only about 15% carbohydrate, as cited by Roos et al. 1975 b). As is found with other glycopeptide hormones, it is composed of two dissimilar subunits each of which, alone, has little or no biological activity although 67% of the biological activity was restored when the subunits were recombined (Reichert et al. 1970). The α-subunit of human LH, comparable to the α-subunit of TSH and of follicle-stimulating hormone, contains 89 amino acids, the N-terminus coinciding with residue 8 of the α-subunit of ovine LH (which can have 96 amino acids and is identical to the α-subunit of bovine TSH). The amino acid sequences have been determined by Sairam et al. (1972) and by Shome and Parlow (1974). The β-subunit contains 115 amino acids with extensive heterogeneity at the carboxyl terminus. Seventy-five of the positions are identical with those of the β-subunit of ovine LH (e. g. Keutmann et al. 1974). The presence of 19–22 cysteine residues and 4–5 methionine residues per molecule (Roos et al. 1975 b) indicates that the molecule could have a variable three-dimensional structure. Nureddin et al. (1972) have shown that it readily undergoes some form of polymerization at certain concentrations of salts or of hydrogen ions.

Roos et al. (1975 b) quoted much work to the effect that when highly purified preparations of human LH are examined by electrophoresis, they show between two and four separate protein components. In their work on the hormone extracted from human pituitaries they found two major and two minor components each of which appeared to be homogeneous when studied by other physicochemical procedures. The biological activity of component I was 1700 ± 300 IU/mg; that of each of the components II and III was 8100 ± 600 IU/mg; while the activity of component IV was 3600 ± 400 IU/mg. Loeber et al. (1977) extracted at least two forms of LH from human pituitary glands. Although their type II LH was three times more potent than type I both in radio-immunoassay and in its ability to inhibit the binding of labelled hormone to rat ovarian tissue, the relative potencies were reversed when these fractions were tested in vivo in the ovarian ascorbic-acid depletion bioassay. These workers pointed out that it is not wise to regard radioligand assays as bioassays, as many investigators do, because the binding in such assays is not necessarily confined to sites which are specific for the expression of biological activity. Certainly their results indicate that there can be marked discrepancies between radioligand and truly biological assays.

II. Methods of Assaying Luteinizing Hormone

The hormone, secreted by the anterior pituitary gland, is said to cause ovulation in the female and to stimulate the production of steroids by the corpus luteum, the latter being possibly related to its ability to increase the 'luteinization' of the ruptured follicles. In the male it stimulates the Leydig cells of the testis to produce androgens and oestrogens (e.g. Nureddin 1977) and, like human chorionic gonadotrophin, it has a direct influence on the accessory reproductive organs (Diczfalusy 1954). Consequently the older in vivo bioassays, which still are of considerable value, depended on one of the following biological responses: the increase in weight of the ventral prostate gland in immature hypophysectomized male rats (Greep et al. 1941); the increase in weight of the seminal vesicle (van Hell et al. 1964); or the depletion of ascorbic acid from the superovulated rat ovary (Parlow 1958, 1961).

From the early 1960s radio-immunoassay has generally replaced these bioassays, having far greater sensitivity and being capable of measuring many more samples in a given time. However, although the radio-immunoassay of LH is very widely used, there is some discontent over its quantitative reproducibility in different laboratories. Even before this became generally recognized, an international collaborative study by several laboratories (Bangham et al. 1973) of a reference preparation (preparation 69/104) of follicle stimulating hormone and luteinizing hormone gave rather discrepant ratios of radio-immunoassay values to bioassay values, when compared to other standard preparations (LER-907 and the 2nd International Reference Preparation). More recently three groups of workers have produced bioassays of luteinizing hormone. The most recent, that of Nureddin (1977) is an in vivo assay in which the standard graph is obtained by injecting the hormone, dissolved in 0.02 *M* sodium borate – 1% NaCl buffer (pH 8.0), subcutaneously into female Wistar rats of 32 days of age (85–100 g in weight). The animals, four for each dose of hormone, are killed 5 h later and the ornithine decarboxylase activity is measured in the ovaries, pooled from all four rats of each group. Human chorionic gonadotrophin (HCG) also stimulates ornithine decarboxylase activity from virtually nothing to high activity at 5 h but whereas the activity induced by LH continued for another 3 h that caused by HCG declined very sharply. The effective dose-range of this assay was 15–40 μg LH/kg. It is too early to assess how valuable this bioassay will prove to be but it deserves mention as a new approach which may be of value.

The other two bioassays are in vitro bioassays done on isolated cells. Dufau et al. (1971) showed that there was a linear relationship between the amount of testosterone produced by the decapsulated rat testes and the dose of HCG or LH applied to them. They then used isolated interstitial cells from the testes and developed a reliable bioassay for LH (Dufau et al. 1974, 1976a, b; Dufau and Catt 1975). Diczfalusy and his group also began with decapsulated testes from mice (Van Damme et al. 1973) and since then have also produced a bioassay depending on the stimulation of secretion of testosterone from Leydig cells isolated from the testes of mice (Van Damme et al. 1974; Romani et al. 1976). The sensitivity of this bioassay varied with the seasonal variation in sensitivity of the cells and was between 3.5 and 8.0 mU/ml plasma (in terms of the 2nd International Reference Preparation); the mean index of precision (230 multiple assays) was 0.040; the weighted coefficient of variance for interassay variation was 7.6%; and the assay gave almost complete recovery of exogenously added hormone with no evidence of either synergistic or antagonistic influence. To obtain parallel dose responses between the standard preparation and the plasma it was essential to have either calf serum or human serum in the assay system. Thus this bioassay (Romani et al. 1977) overcomes the discrepancy between this form of bioassay and radio-immunoassay which had previously been reported by Quazi et al. (1974). However, although the biological activity and the immunological activity showed similar changes during the menstrual cycle, there appeared to be 5.5 times the amount of bioreactive LH compared to immunoreactive LH taken over the whole cycle; this may be due to injudicious selection of the standards used in these two types of assay (Romani et al. 1977).

The isolated interstitial cell bioassay of Dufau and Catt (1975; Dufau et al. 1976 b) was done in the presence of 1-methyl-3-isobutyl xanthine. It also required the presence of gonadotrophin-free serum, or 5% bovine serum albumin, to maintain a constant proportion of serum protein in both the standard and the test samples. It is normally done on serum. The within-assay coefficient of variation for pooled normal male plasma containing 30 mU/ml was $\pm 10\%$; the between-assay variation was $\pm 15\%$. The index of precision was 0.035 ± 0.014 (n, 76). In 42 normal females the bioassay: immunoassay ratio was 1.2 ± 0.4 with no consistent change during the menstrual cycle; higher ratios were observed in normal males and in post-menopausal females. The limit of detection is said to be 50 μU for hMG and 20 μU for hCG. In 42 normal females the follicular and luteal phases assayed at 29 ± 16 mU/ml; in three females at mid-cycle the value was 189 ± 59 mU/ml; the concentration of LH in ten post-menopausal females was 223 ± 92 mU/ml.

Thus both these methods (Dufau et al. 1976 b; Romani et al. 1976) seem to be eminently suitable for the assay of the biological activity of LH in plasma.

B. Cytochemical Bioassay of Luteinizing Hormone

I. Original Procedure

1. Background

Before the isolated-cell bioassays had been fully developed, there was a need for a procedure which would readily measure the biological activity of LH in individuals. Rees et al. (1973 a; Kramer et al. 1974) therefore exploited the fact that LH depletes ascorbate from the superovulated ovary (Parlow 1961), very much as ACTH depletes ascorbate from the adrenal gland. Basically, therefore, they applied the same technique to the ovary as had been developed for the first cytochemical bioassay, namely that for ACTH (as discussed in Chap. 7).

2. Original Cytochemical Segment Bioassay
(Rees et al. 1973 a; Kramer et al. 1974)

Because the assay was a within-animal assay, it was helpful to have the ovaries as large as possible, preferably uniformally filled with luteinized cells. Hence the female weanling albino Wistar rats (of 21–25 days of age) were injected with 50 IU of pregnant mare's serum, followed 50–60 h later by 50 IU of human chorionic gonadotrophin to produce the superovulated ovaries of greatest weight and greatest degree of luteinization. The rats were killed 112 h later (Kramer et al. 1974).

The ovaries were removed and cut into suitable segments (normally three segments from each ovary); these segments were maintained individually for 5 h at 37° C in Trowell's T8 medium with $10^{-3} M$ ascorbate (as in the cytochemical bioassay of ACTH: Chap. 7). Each of four of the segments was then exposed to one of a series of graded concentrations of a standard preparation of the hormone (0.005–50 mU/ml; approximately 1–100 pg/ml); two were exposed to one of two concentrations of the first plasma to be assayed, at either 1 : 10 or 1 : 100 dilution; two more with the second sample and so on. Sometimes it was possible to assay three samples, each at two dilutions, and have four segments for the standard calibration graph, all on tissue from the same rat.

The sensitivity of the assay varied with the variety of rat used, and even with different rats of the same strain. However, because all the tissue used in any one group of assays came from the same animal, differences in response by different rats did not seriously influence this assay. The sensitivity was also dependent on the standard preparation of LH, the MRC LH 68/40 preparation giving improved sensitivity over that used in the first study by Rees et al. (1973 a).

After the segments had been exposed to the standard preparation of the hormone, or to the diluted plasma samples, for 4 min, they were chilled, sectioned at 12 μm and reacted for ascorbate by the Prussian blue reaction, as is done for the cytochemical bioassay of ACTH (Chap. 7). The amount of ascorbate left in the cells was measured by microdensitometry (as discussed for ACTH: Chap. 7) although Kramer et al. (1974) measured only at the absorption maximum of the dye (680 nm) and found it was unnecessary to subtract the values at 480 nm, apparently because the non-specific absorption (at 480 nm) due to this tissue was negligible.

With the MRC 68/40 reference preparation as the standard, the response was linear over the range 0.005–5.0 mU/ml, allowing the measurement of 0.5 mU of LH in 1 ml of plasma (Holdaway et al. 1974 a). The mean index of precision, in 16 consecutive assays done by two workers, was 0.12. The specificity of the assay was confirmed (Rees et al. 1973 a) against purified hFSH, ovine prolactin, ACTH (3rd International Working Standard and synthetic α_{1-24} ACTH), synthetic arginine vasopressin and purified hTSH.

The α- and β-subunits of LH, tested separately, gave less than 10% of the activity of the whole molecule, but they gave full activity after they had been recombined in 0.05 *M* phosphate buffer, pH 7.5, for 78 h at 4° C (Holdaway et al. 1974 a). However, the standard dose-response graph was severely altered when LH was tested in the presence of 10 ng/ml of the α-subunit, and more so with 10 ng/ml of the β-subunit. Follicle stimulating hormone, when tested alone, cross-reacted only slightly in this assay system; it had no appreciable influence on the standard dose-response to LH if present at a concentration of 1 ng/ml or less but markedly lowered the activity when the concentration of FSH was increased to 10 ng/ml (Kramer et al. 1976).

Dilutions of plasma gave dose-response curves which were parallel to those induced by the standard preparation in 88% of the samples tested. Plasma concentrations in normal males and females were in good agreement with values obtained by radio-immunoassay but there were discrepancies in the results of this assay and of radio-immunoassay in assaying higher levels of LH, as in plasma from females at mid-cycle and subjects who had received an injection of gonadotrophin-releasing hormone. The latter discrepancies could have been due to a true dissociation of biological and immunological reactivity in the hormone in the circulation of the subjects examined, particularly because the subjects were sampled only once after the injection. (Such true discrepancies, due to different half-times of the biologically active hormone and its immunoactivity, have been elucidated for TSH and for ACTH: Chaps.10 and 7.)

3. Discrepancy Between Results with this Bioassay and with Radio-immunoassay

The main drawback to the use of this bioassay was the finding of Holdaway et al. (1974 a; fully confirmed in a later study by Kramer et al. 1976) that in the nine normal menstrual cycles studied (in the two communications), the peaks of concentration of LH, measured by the cytochemical assay, did not coincide with the well-known immunoreactive peak (at ovulation, referred to, as Day 0 of the cycle). Thus whereas radio-immunoassay gave a sharp peak, equivalent to a concentration of about 40 mU/ml, at Day 0, the bioassay gave a peak (about 45 mU/ml) at Day −2, a low level (about 8 mU/ml) at Day 1, and a second peak (about 40 mU/ml) at Day 4.

The fact that this bioassay gave preovulatory and luteal peaks, with an apparent trough between the two, is puzzling. One possibility, currently being explored with the new section assay (see below), is that this represents a quirk of the cells of the superovulated ovary. It may well be objected that this possibility is unlikely in view of the fact that one of the best bioassays of LH, the Parlow ovarian ascorbic acid depletion assay (Parlow 1958, 1961), utilizes superovulated ovaries. However, it must be appreciated that the cytochemical assay can also produce results equivalent to those found by radio-immunoassay (Holdaway et al. 1974 a); it is only when samples are taken from individuals during the menstrual cycle that the differences become glaringly difficult to explain. Moreover, other measurements of bioactive LH in plasma at different times of the cycle, done by other bioassay methods (Watson 1972), have shown a broader peak of LH activity around mid-cycle than is normally seen by radio-immunoassay. A somewhat similarly broad excretion of urinary LH, measured by a bioassay, has also been reported (Fukushima et al. 1964; Brown et al. 1964, reported multiple peaks of LH bioactivity in human urine). Indeed the sharp peak at mid-cycle, so dear to immunoassayists, is by no means as certain for biologically active LH, as was shown in the studies on bovine serum collected during complete cycles in which two or three peaks of bioactive or immunoreactive LH were detected by Schams and Karg (1969) and by Snook et al. (1971).

For all that, in the normal human female, radio-immunoassay invariably demonstrates a sharp rise and an equally sharp fall in the circulating level of LH just at mid-cycle. In contrast, this cytochemical bioassay showed a pre-ovulatory and a luteal peak, with a low concentration of LH at mid-cycle. It is possible, as suggested by Kramer et al. (1976), that this discrepancy is due to the presence of some unidentified interfering material in the plasma at mid-cycle, although these workers considered that this interference was unlikely to be caused by FSH or by the separate subunits of LH which never reach circulating levels that could cause such interference. On the other hand, this strange discrepancy could have a more fundamental, and potentially valuable explanation:

It has been argued (Chap. 7) that the depletion of ascorbate from the adrenal gland is an essential step in steroidogenesis in cells, such as those of the zona reticularis, in which the mechanisms for producing steroids are held in abeyance by the lack of available oxygen for hydroxylation by cytochrome P-450. If the cells of the superovulated ovary have a similar biochemical mechanism, the depletion of ascorbate by LH must be considered to be an essential part of steroidogenesis by these cells. If that is conceded, it is not surprising that the cytochemical bioassay of Rees et al. (1973 a) shows peaks of steroidogenic LH during the pre-ovulatory and luteal phases, namely when the cells are expected to be producing steroids and therefore require hydroxylation by the cytochrome P-450 system. Thus the results of this cytochemical bioassay are consistent with the steroidogenic activity of LH. The real question with which we are left is what is the biological significance of the immunoreactive LH peak just at the time of ovulation? Indeed, when it is put this way, we are left wondering how anyone could have expected a steroidogenic influence to be anything other than minimal at mid cycle. This speculation leads to the possibility that what we generally call 'luteinizing hormone' in the female has two quite disparate functions: during the pre-ovulatory and luteal phases it functions to assist steroidogenesis, but at mid-cycle it is altered to influence ovulation. If this is correct, then the mid-cycle form of bioreactive LH must be measured by its biological, or biochemical, influence on ovulation, not on steroidogenesis (as evinced by the depletion of ascorbate which, as far as I am aware, cannot induce ovulation).

Thus this hypothesis requires that the molecule which we call luteinizing hormone can occur in two different configurations. The best-known configuration would be that which is recognized by radio-immunoassay, the biological action of which is concerned with ovulation. It follows that the biological activity of this form can be assayed only by some influence which is associated with ovulation. The second form, which has a lower immunopotency, is concerned with steroidogenesis and can be assayed by its ability to deplete cells of ascorbate. There is some evidence in support of this hypothesis. Loeber et al. (1977) isolated two forms of LH: LH type II was three times more potent than LH type I as measured by radio-immunoassay but the reverse potencies were found with the ovarian ascorbic acid depletion bioassay. Roos et al. (1975 b) found somewhat comparable results (as discussed in Chap. 11.A.I.) in fractions which apparently had similar molecular weights but which could be separated electrophoretically, implying that they may have had different configurations. If a cytochemical bioassay can be developed based on a biochemical component of the ovulatory process, it will be interesting to see whether some of these fractions assay predominantly in such a system whereas others have their main potency in an ascorbate-depletion assay.

It may be remarked that there is some indication that FSH may be pleomorphic. Diebel et al. (1973) found evidence for steroid-regulated pleomorphism of this hormone in the rat. Peckham et al. (1973) used exclusion chromatography to show that certain preparations of FSH from the pituitary glands of rhesus monkeys were heterogeneous. The ratio of immunoassayable to bioassayable activity was least for the largest molecules of FSH. There was a preponderance of the larger molecular weight FSH in extracts and sera from ovariectomized monkeys. However, differences in sialation could have accounted for some of these results, particularly those differences which they observed in the half-time of these molecules in the circulation. Similarly the different clearance times of the different types of FSH in rats which had been subjected to low concentrations of either androgen or oestrogen after orchidectomy (Bogdanove et al. 1974) may also only reflect different degrees of sialation. For all that, Bogdanove et al. (1974) considered that androgens altered the type of FSH secreted. Thus it is possible that FSH is pleomorphic, with the larger molecular weight form having a markedly different bioassay: immunoassay ratio. We have seen (above) that LH too appears to show pleomorphism. It is therefore not beyond possibility that the pleomorphism of LH may even extend to the degree required by this speculation.

However, it must be emphasized that this hypothesis does not explain why the isolated-cell bioassays, based on the LH-induced production of testosterone by isolated interstitial cells of either the rat or the mouse, do not appear to show this discrepancy. Admittedly, interstitial cells are steroidogenic; they have no part in ovulatory mechanisms and presumably could not bind a specifically ovulatory configuration of LH even if the hormone molecule ever assumed such a configuration.

II. Cytochemical Section Assay

1. Background

The results of the segment bioassay of Rees et al. (1973 a) were sufficiently impressive that my colleagues and I felt it worth reinvestigating this method, especially with a view to producing a section assay, with all the advantages of such assays (as discussed in Chap. 8). One of the advantages of a section assay is that it can be done on a single corpus luteum; consequently it was not necessary to have large ovaries as was required for the segment assay where one segment of the whole ovary was needed for each point in the assay: the larger the ovary, the more samples that could be measured in each within-animal assay. The use of sections allowed us to avoid using superovulated ovaries, which cannot be considered to be physiological.

We were fortunate in having Professor Judith Weisz join us in this work. This was made possible by a grant from the Medical Research Council: tribute must be paid to the scheme, operated by the Council, by which senior workers with expert knowledge, such as Professor Weisz, are assisted to visit workers in other laboratories for collaborative studies where each contributes his or her particular expertise. She pointed out that possibly the most hazardous step in the assay procedure, as developed previously, was the maintenance culture of ovarian segments. One of the reasons for maintaining hormonal target tissue in vitro for 5 h is to remove it from the hormonal influence of the animal and she pointed out that on the second day of dioestrous, and on the 4th day of pregnancy, the corpus luteum was exposed to very low circulating levels of LH: in a sense, Nature had done the in vitro maintenance for us.

Initial experiments showed that sections could be made to respond to low concentrations of LH. The problem was that when serial sections were immersed for the same time in the vehicle used for diluting either the standard preparation of hormone or the sample of plasma, the results were very variable. Sometimes the vehicle produced no change in the concentration of ascorbate in the cells, as was expected; sometimes the vehicle appeared to cause as much depletion of ascorbate as did the lower concentrations of the standard preparation of the hormone used in preparing the standard calibration graph (Fig. 11.1). In other work there had been some indication that inexplicable results can be obtained if cells are not under some degree, however slight, of hormonal control. It was therefore decided to prime the tissue with a very low concentration of luteinizing hormone, preferably one-tenth of the lowest concentration that could be detected readily by the ascorbate-depletion test. The result of such priming has been that sections exposed to the vehicle alone now invariably show no depletion of ascorbate (or of reducing potency).

Another valuable suggestion was to use Sprague-Dawley rats which can have seven ovulable follicles in each ovary at any one time, both ovaries being simultaneously functional.

2. Procedure (Buckingham et al. 1979)

Sprague-Dawley rats of about 250–300 g are killed by cervical dislocation either during the 2nd day of dioestrous or on the 4th day of pregnancy. Both give very comparable results; the choice depends mainly on availability although the latter may give more sensitive responses. The ovaries are removed and deftly dissected free from peri-ovarian tissue. They are then placed on defatted lens tissue on a metal mesh table in a vitreosil dish, very much as for the in vitro maintenance culture, and Trowell's T8 culture medium, containing 10^{-7} IU of LH per ml, is poured over them. They are left in contact with this priming medium for 4 min and then they are chilled to $-70°$ C in hexane for up to 1 min (as described previously). The tissue is stored at $-70°$ C but must be used within 48 h of chilling.

Sections are cut to determine the presence of a suitable corpus luteum: for this purpose they can be stained with toluidine blue (e. g. Chayen et al. 1973 b) to identify the large corpus luteum of the latest generation of these structures. (They also lack 20 α-hydroxysteroid dehydrogenase activity, in contrast with those of previous generations.) Subsequently, serial sections are cut at 20 µm and clipped in duplicate, back to back, in the same apparatus as is used for the cytochemical section assay of

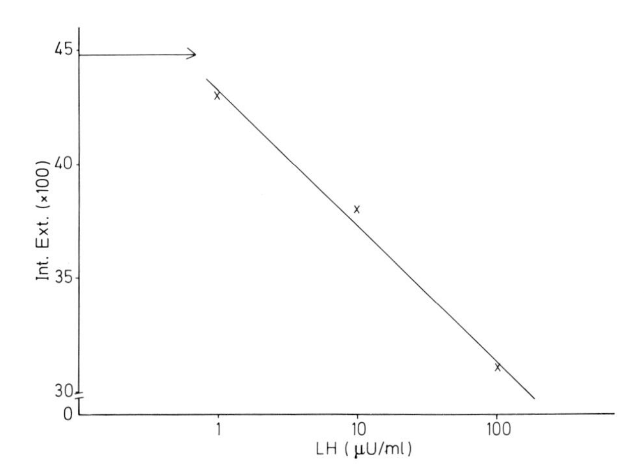

Fig. 11.1. The negative linear relationship between the concentration of LH applied to sections and the amount of reducing potency (ascorbate) retained in the sections (Int. Ext. × 100). The reducing potency in other sections, exposed for the same time to the vehicle alone, is shown by the *arrow*. Thus increasing concentrations of the hormone caused increasing depletion of ascorbate

ACTH (Chap. 8) or of TSH (Chap. 10). Each of the first four compartments, each with two sections, is filled with one of a graded series of concentrations of the standard preparation of the hormone (the International Reference Standard 68/40 or 69/104 has been used in developing this method), ranging from 10^{-7} to 10^{-4} (inclusive) IU/ml. The hormone preparation (or the sample of plasma) is diluted in Trowell's T8 medium containing 10^{-3} M ascorbate, with 5% of the colloid stabilizer Polypeptide 5115, (Sigma) at pH 7.6. The next two compartments contain the first plasma to be tested, usually diluted 1:100 and 1:1000 respectively; the next two compartments contain similar dilutions of the second plasma, and so on. Including the eight sections needed for the calibration graph (Fig. 11.1) up to five samples can be assayed simultaneously on one corpus luteum. More can be done on each segment if two suitable corpora lutea are present.

The sections are exposed for 80 s (Fig. 11.2) at 37° C, the whole apparatus, the sections and the solutions having previously been brought to this temperature. The rack holding the sections is then removed and placed in the bath which contains the ferric-ferricyanide reagent for demonstrating reducing potency (as in Chap. 8; also Appendix 3). Normally 9 min in this reagent (three freshly prepared baths, each for 3 min) produce sufficient reaction; it may be necessary to use three periods of 5 min, each in freshly prepared reagent. The sections are rinsed, dried and mounted. The amount of reaction product, equivalent to the amount of reducing groups (mainly ascorbate), present in virtually the whole corpus luteum is measured in 10–20 fields if a × 10 objective is used with a large mask (C1 on the Vickers M85 microdensitometer) and a small scanning spot (spot 1). Measurements are made at 680 nm only.

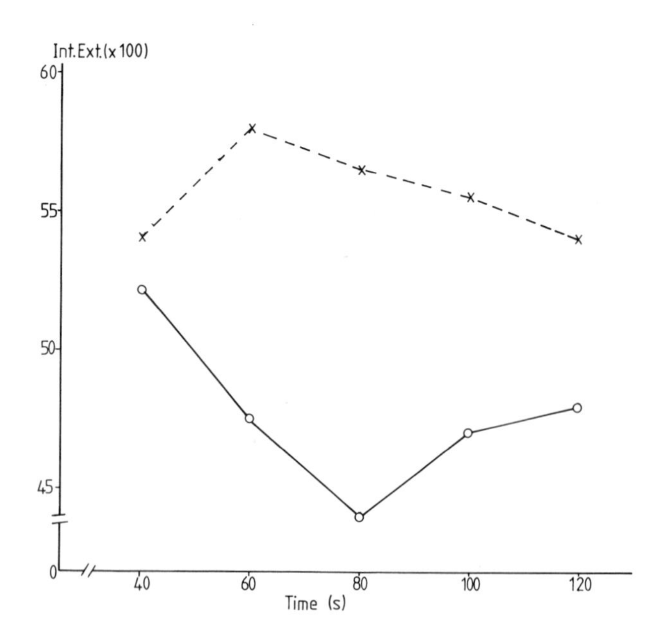

Fig. 11.2. The time course of ascorbate depletion (loss of ability to produce the Prussian blue reaction) and hence loss of extinction (Int. Ext. × 100) in sections treated for various times with LH (100 μU/ml), shown by the *solid line*. The *broken line* indicates the effect of vehicle alone acting on serial sections

3. Validation

Luteinizing hormone depletes ascorbate from the corpus luteum. Consequently the assay shows an inverse linear relationship between the amount of the reaction product for reducing groups and the logarithm of the concentration of the hormone, from 10^{-7} to 10^{-4} IU/ml. The dose-responses of serial dilutions of human and of rat plasma are parallel to those of the standard preparations used (International Reference Standard 68/40 and 69/104) although non-parallel responses have been obtained with plasma from chickens and owl monkeys.

The mean index of precision in ten assays was 0.12 ± 0.01. Of many steroids and pituitary peptides tested up to now, only TSH and FSH, in concentrations 100–1000 times the circulating levels found in the human under normal physiological conditions, affected the ovarian responses to LH. Both in human and rat plasma the method has yielded results which were in reasonable agreement with those obtained for the same samples by radio-immunoassay. At the time of writing, the validity of the method is still being tested (Buckingham et al. to be published). It is yet too early to be certain whether this assay will, or will not, agree with radio-immunoassay over the mid-cycle peak. Either way, it should provide a useful tool for examining the mode of action of the hormone we call luteinizing hormone.

Chapter 12

Gastrin-like Activity

A. Background

I. Nature of Gastrin

Gastrin, or little gastrin, is a heptadecapeptide amide, of molecular weight about 2100, with a single tyrosine residue which may (as in G_{17} II) or may not be (G_{17} I) sulphated (Gregory and Tracy 1961, 1964; Gregory et al. 1966; also see Gregory 1969). It can occur as big gastrin (Yalow and Berson 1970, 1971), which has 34 amino acids and also can be isolated in the sulphated or non-sulphated condition, or even as big big gastrin (Yalow and Berson 1972). Berson and Yalow (1971) found the heptadecapeptide (G_{17}-gastrin) to be the major component in extracts of antral mucosa but the proportion of big gastrin (G_{34}) increased distally in the digestive tract, being the sole detectable component in extracts of the mucosa of the proximal jejunum. Both components were immunologically indistinguishable. A 'mini gastrin', of perhaps 13 amino acids (G_{13}), equivalent to the 5–17 sequence of little gastrin (G_{17}-gastrin), or to the 21–34 sequence of big gastrin, and capable of stimulating the secretion of gastric acid, has also been reported in tumour tissue (Gregory and Tracy 1974). The different larger forms of the gastrin molecule have been reviewed by Yalow (1974).

The N-terminal portion is said to be biologically inactive (Dockray and Walsh 1975) although there is recent evidence (Chap. 15) that this 'inactive tail' does influence the selectivity of the response of the target cells. Of the whole of the G_{34}-gastrin molecule, only the last four amino acids of the carboxy-terminus (Numbers 34, 33, 32 and 31: phenylalanine, aspartate, methionine and tryptophane) seem to be essential for biological activity (Tracy and Gregory 1964): the commercially available 'pentagastrin' (Peptavlon) appears to have full biological potency for gastrin activity. The smaller 'gastrins', G_{13} and G_{17}, represent cleavage of the peptide sequence of G_{34}-gastrin at the appropriate positions starting from the carboxy-terminus. They both contain the biologically active region of the gastrin peptide. It is noteworthy that the first five amino acids of the carboxy-terminus of cholecystokinin (CCK) are identical with the equivalent amino acids of gastrin; there has been understandable concern that the octapeptide of cholecystokinin (CCK-OP), which could well be expected to be present in the circulation, should show biological and immunological properties practically identical with those of the biologically active part of the gastrin polypeptide. Equally the whole CCK molecule, having these amino acids at its carboxy-terminus, could mimic gastrin.

Although radio-immunoassays have been developed for many gastrointestinal hormones, Walsh and Grossman (1975) concluded that 'because of the heterogeneity of circulating gastrin and because different molecular forms vary in biologic activity, total gastrin activity determined by radio-immunoassay cannot be more than a crude index of bioactivity.' The more recent radio-immunoassay which is said to be specific for the heptadecapeptide gastrin (Dockray and Taylor 1976) may improve this situation.

II. Biological Effects of Gastrin

Gastrin is produced by gastrin cells (G cells) present in the gastric antrum and in the duodenum. It has also been shown, by immunofluorescence, to be present in certain cells of the pancreas. The immunofluorescence and the immunoperoxidase procedures have been exploited to locate many diverse gastrointestinal 'hormones', or potential hormones, in cells in various tissues of the gastrointestinal tract. Particularly in view of the structural similarities of parts of the molecules of different known gastrointestinal hormones, such immunological staining procedures require very careful and detailed tests for specificity, quite apart from the danger, well known to histochemists, of apparently specific adsorption of specific molecules to non-specific sites.

In man the half-life of natural G_{17}-gastrin, measured by radio-immunoassay, was only 5 min; that of the G_{34}-gastrin was 42 min (Walsh et al. 1975). These values may be greater than the half-lives of the biologically active molecules because, for other polypeptide hormones, immunoreactivity can be measured in the circulation long after bioreactivity has been lost. Maximal rates of acid secretion by G_{17} and pentagastrin were similar; big gastrin caused lower but prolonged stimulation. The normal circulating level of immumoreactive gastrin, in man after an overnight fast, is about 24 pmol/litre (Stadil and Rehfeld, 1973). Gastrin is believed to be one of the factors involved in the stimulation of gastric acid secretion; it appears to stimulate gastric motility and may influence the gastric clearance time; it also acts as a trophic hormone, stimulating the growth of the gastrointestinal mucosa (e.g. Johnson 1975). The bioassays of gastrin, done in intact animals or on isolated gastric mucosa, have depended on the stimulation of gastric acid secretion. The limit of sensitivity of the bioassay devised by Blair and Wood (1968) was 25 pmol; given 50 ml of plasma they could assay concentrations greater than 0.5 pmol/ml (or 1 ng/ml).

B. Cytochemical Segment Assay

I. Carbonic Anhydrase

1. Background

As with the other bioassays, the cytochemical bioassays have depended on the fact that gastrin stimulates the secretion of gastric acid. Thus to produce a cytochemical bioassay it is necessary to select a critical biochemical event in the chain of metabolic processes that produce the final physiological phenomenon by which the hormone is characterized. Unfortunately, the mechanism by which the parietal cells of the gastric fundus secrete acid, which is the final physiological phenomenon most characteristic of gastrin-like activity, is not clearly understood, but there seems little doubt that it must involve the activity of carbonic anhydrase (Salganik et al. 1972; Narumi and Kanno 1973).

This enzyme (E.C. 4.2.1.1., also called carbonic dihydratase) has been much studied by Maren (1967). It is concerned with the maintenance of the equilibrium between carbon dioxide and carbonic acid, and is found in secretory organs that are concerned with the accumulation or transfer of either hydrogen ions or bicarbonate ions (also see Davenport 1962; Bittar 1964). It has a molecular weight of 34 000, with one zinc atom per molecule. If this zinc is replaced (except by cobalt) or complexed, the enzyme becomes inactivated (Coleman 1967). Sulphonilamides and particularly acetazolamide are inhibitory. The enzyme is concerned in the equilibrium between carbon dioxide and bicarbonate ions, i.e. $HCO_3^- \rightleftharpoons CO_2 + OH^-$. Thus it can be measured either by the rate of evolution of carbon dioxide, as it is in biochemical methods, or by the generation of hydroxyl ions. The latter was used by Hansson (1967) for a histochemical method (based on that of Häusler 1958) for demonstrating the activity of this enzyme.

2. Cytochemical Reaction

In the method developed by Loveridge (1978) the cryostat sections (18 μm) are first equilibrated to room temperature for 10 min; this helps to retain them on the slides during the subsequent, rather vigorous, conditions. During that time, the following reaction solutions are prepared:

Solution A: 0.1 *M* cobaltous sulphate 6 ml
0.5 *M* sulphuric acid 6 ml
0.067 *M* potassium dihydrogen orthophosphate 1–3 ml
This solution is made up to 17 ml with distilled water.

Solution B: 0.75 g sodium hydrogen carbonate dissolved in 0.1 *M* HEPES buffer, adjusted to pH 7.4 with sodium hydroxide, and containing 0.005% gum tragacanth.

Equal volumes of these solutions are mixed immediately before use and allowed to stand for 2 min before being poured into a flat trough which contains the sections lying horizontally. The base of the trough has spacers for holding the slides. The volume of the final reaction medium (solutions A + B), required to cover the sections by 0.8 mm, is calculated from the area of the trough and preferably by direct measurement, using blank slides to compensate for the volume which will be occupied by the sections on slides. The reaction is allowed to proceed for 1–3 min with the trough being continually but gently agitated.

The sections are then washed in tap water, immersed in a saturated solution of hydrogen sulphide and mounted in the water-soluble Farrants' medium.

3. Rationale

Following Maren (1967), the main reaction mediated by this enzyme is $H_2CO_3 \rightleftharpoons CO_2 + H_2O$. This, however, involves subsidiary reactions, the most important of which (Hansson, 1967) are as follows:

$$H_2CO_3 \rightleftharpoons CO_2 + H_2O,$$
$$H_2CO_3 \rightleftharpoons HCO_3^- + H^+,$$
$$HCO_3^- \rightleftharpoons CO_2 + OH^-.$$

Thus bicarbonate with sulphuric acid is the substrate (solutions B + A) yielding free CO_2 which forms HCO_3^-. The enzyme liberates CO_2 and hydroxyl ions; the latter are trapped by the cobalt ions (solution A) to form insoluble cobalt hydroxide, which is later converted into the brown or black cobalt sulphide. However, since CO_2 is an end product of the reaction, it is inhibitory and so must be removed quickly. Hence the need to ensure that the sections are not more than 0.8 mm from the surface of the reaction medium and also that the reaction medium should be gently agitated during the reaction. Phosphate is included in the reaction medium because it enhances the precipitation of the reaction product. Whether this is due to the formation of a more insoluble cobalt-phosphate-hydroxide complex, as suggested by Hansson (1967) or whether it is due to activation of an isoenzyme, carbonic anhydrase B (Christiansen and Magid 1970), as seems likely (Loveridge 1978), or to a combination of these factors, is not completely clear.

When solutions A and B were first mixed, the pH was about 6.2 but rose to 6.8 after 2 min. Provided that the concentration of the HEPES buffer was 0.1 *M*, the pH rose only very slowly, reaching pH 7.3 3 min later. It was still below pH 7.4 after the solutions had been mixed for 10 min. Thus the cytochemical reaction is done at pH values ranging from 6.8 to 7.3 (Loveridge 1978).

The purpose of the low concentration of gum tragacanth is to retard the deposition on the sections of uncatalysed reaction products from the reaction medium. These become serious at longer times of the reaction: should longer times be deemed necessary, the medium should be replaced every 3 min by fresh medium which has already been allowed to reach pH 6.8 before it is used. The reaction was linear with increasing thickness of sections from 10 to 22 μm, and it was linear with time over the first 4–5 min. Acetazolamide at $10^{-5} M$ caused 56% inhibition of the activity (Loveridge 1978).

4. Measurement

Because the absorption curve of the reaction product is broad (Loveridge, 1978), the reaction product is measured at 550 nm, which allows the greatest sensitivity of most photomultipliers. It is measured with a Vickers M85 scanning and integrating microdensitometer with a mask (20 μm diameter) which is just filled by one parietal cell (magnification $\times 20$). A small spot size (size 1) is normally used. Thus 10 or 20 parietal cells in each of two duplicate sections are measured. Then, to correct for light scattering by the tissue and especially for non-specific adsorption of cobalt, a similar number of fields are measured in the nearby muscle; the mean of these readings is subtracted from that of the parietal cells. By this technique, the variation in the activity measured in 20 serial sections was $\pm 6\%$ (Loveridge 1978).

II. Segment Assay (Loveridge et al. 1974; Loveridge 1977)

1. Background: pH of the Medium

As in all the cytochemical segment assays, the target tissue is maintained in vitro for 5 h to remove the responsive cells from the influence of the hormone endogenous to the guinea-pig and to allow them to recover from the trauma associated with the death of the animal and with excision of the tissue (as discussed in Chap. 4). It was found, however, that the pH of the culture medium greatly affected the carbonic anhydrase activity of the parietal cells, lower pH values giving lower activity. Consequently it was advisable to maintain the tissue at as low a pH as was consistent with maintaining satisfactory histological appearance of the tissue, namely at pH 7.0. Various conditions were investigated (Loveridge 1977) including variation in the proportion of CO_2 in the gaseous phase. The best buffering was obtained by decreasing the concentration of bicarbonate in the Trowell's T8 medium and using the normal 5% CO_2 : 95% O_2 mixture for the gaseous phase. The required concentration of bicarbonate was achieved by adding 0.2 ml of 1-N-hydrochloric acid to 10 ml of Trowell's T8 medium and then passing 95% O_2 : 5% CO_2 gas through this to drive off the excess of carbon dioxide generated by the action of the acid on the bicarbonate in the normal medium.

2. Procedure

Segments of guinea-pig fundus are removed from a guinea-pig (Hartley strain; weight 300–600 g; sex immaterial) which has been killed by asphyxiation with nitrogen. The strips are cleaned of debris: four are then maintained separately in Trowell's maintenance culture (see Chap. 4) at 37° C for 5 h over Trowell's T8 medium adjusted to pH 7.0, with 95% O_2 : 5% CO_2 as the gaseous phase. These will provide the calibration graph. Two others are maintained similarly for ultimately testing the effect of one plasma, at 1 : 100 and 1 : 1000 dilutions; others may be maintained for testing other plasma samples. At the end of the 5 h the medium is replaced by T8 medium (at pH 7.0) containing either one of four graded concentrations of the standard preparation of the hormone, from 5 fg/ml to 5 pg/ml (the first four segments) or one of two concentrations of the plasma that is to be assayed. The hormone is allowed to act for 5 min, after which the segments are chilled to −70° C. They may be stored at this temperature for up to a week.

Sections are cut at 18 μm and reacted for carbonic anhydrase activity (as discussed above). The activity in the parietal cells in duplicate sections is measured by microdensitometry. In the earlier assays, commercially available pentagastrin was used as the standard reference preparation; in later work, the G_{17}-gastrin MRC Standard was used. They were shown to be equipotent on a molar basis.

3. Precision

Under these conditions, exposure to gastrin caused a rise in the activity of carbonic anhydrase which was maximal at 5 min; further exposure to the hormone showed diminished activity (Fig. 12.1) The dose-response was linear over the range 5 fg/ml to 5 pg/ml and was parallel to the activity induced by the dilutions of plasma (Fig. 12.2). Acetazolamide, at 10^{-5} *M*, abolished the rise in activity recorded otherwise in the presence of these concentrations of the standard preparation, or of dilutions of plasma.

The mean coefficient of variation in activity in four serial sections from segments treated with 5 fg/ml, 50 fg/ml, 500 fg/ml and 5 pg/ml, in four separate studies, was 3.6%, the greatest variation being 8.1%. The variability was unrelated to the concentration of gastrin applied.

In 36 assays the mean index of precision (λ) was 0.11 with only six having a λ of greater than 0.2, the higher values being obtained when higher concentrations of gastrin (up to 50 pg/ml) were assayed. The mean slope was 3.2 (i. e. an increase of 3.2 units of mean integrated extinction multiplied by 100 for each logarithmic increase in dose). The fiducial limits (P, 0.95) varied from 71.6%–139.8% to 97.6%–102.9%. In some specimens the response to dilutions of plasma was not parallel to the dose-response produced by the standard preparation of gastrin (i. e. showing more than

$\pm 15\%$ deviation), implying that these samples contained active moieties other than gastrin (see below). Recovery of gastrin added to two previously assayed samples of plasma was 87% and 119%.

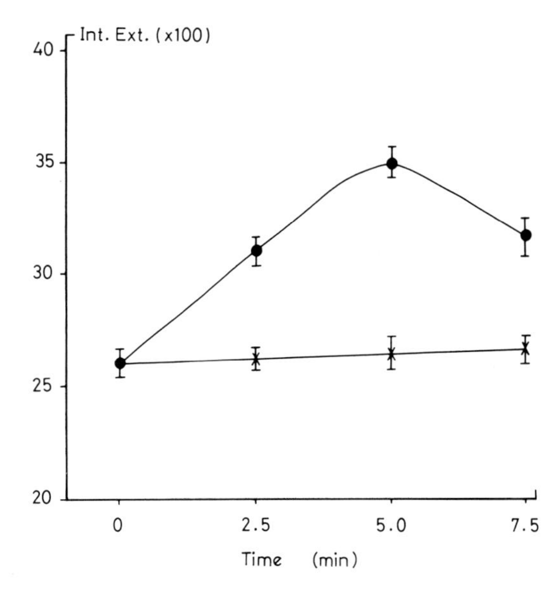

Fig. 12.1. The change in carbonic anhydrase activity (Int. Ext. × 100) with time of exposure to gastrin (*upper line* and *circles*) (G_{17}-gastrin: 5 pg/ml). The *crosses* show that the activity, in serial sections exposed to the vehicle alone, did not change over the course of this period. The bars show the mean activity in each of the duplicate sections

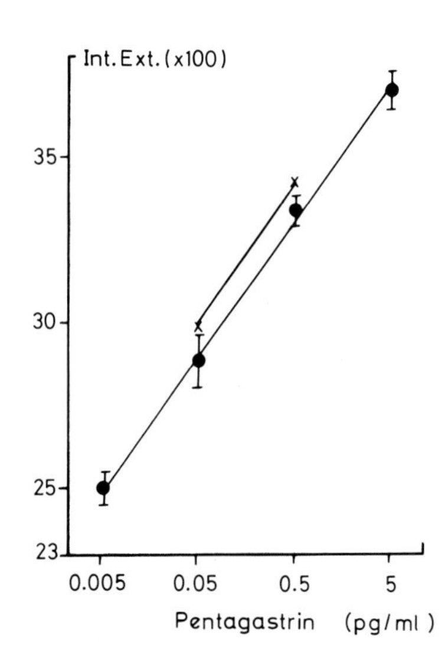

Fig. 12.2. Relationship between carbonic anhydrase activity (Int. Ext. × 100) and the concentration of Pentagastrin applied to sections (*solid circles*). The linear activity/log-dose relationship is paralleled by the activity induced by a sample of plasma diluted $1:10^2$ and $1:10^3$ (*crosses*)

C. Cytochemical Section Assay

I. Problems

It was first demonstrated that sections, of suitable thickness, of the guinea-pig fundus would respond to gastrin in the same way that the segments did in the earlier segment assay. In utilizing this effect for developing a section assay, the first problem was this: In the other section assays the sections, mounted back to back in duplicate, are exposed to the hormone by immersing them vertically in separate compartments of a trough; after the requisite exposure time they are moved, together, to the cytochemical reaction trough in which they are also held vertically. This had to be altered, in view of the fact that the cytochemical reaction for the bioassay of gastrin was the demonstration of carbonic anhydrase activity, for which the sections had to be laid horizontally under a known depth of reaction medium (Chap. 12., Sect. B.I.2.). It was not practicable to expose the sections to the hormone while they were held vertically, and then frantically to unclip them and lay them horizontally before adding the cytochemical reaction medium; it would involve an indeterminate time interval, during which the hormone would still be acting. It was therefore decided to expose the sections to the hormone while the sections were lying horizontally, in the same trough as would then be used for the cytochemical reaction. To ensure that each section was exposed to its relevant concentration of the standard preparation of the hormone, or of the plasma, it was found convenient to have a small syringe mounted above each section. Thus each syringe is filled with the relevant concentration of the hormone or plasma; the sections are placed in the grooved base of the vessel so that each section is just below the relevant syringe; then all the syringes are emptied by means of a handle which links the syringes together. In the apparatus shown in Fig. 12.3, four graded concentrations of a standard preparation of gastrin and five samples of plasma, each at two dilutions, can be tested simultaneously and in duplicate. This means that the whole procedure, from cutting

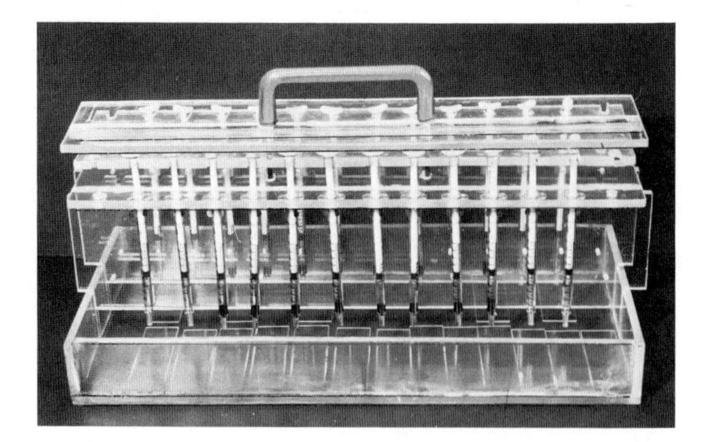

Fig. 12.3. The apparatus used for the section assay of gastrin-like activity. The sections, on slides, are placed between ridges in the base of the trough in which they will later be reacted for carbonic anhydrase activity. Above each of the sections there is a 1 ml syringe, mounted in a head-piece. The handles of all the syringes are set together in a Perspex plate with a handle on top, so that all can be emptied synchronously

the sections to the final result, can take 4 h, and in this time five samples can be assayed (Hoile and Loveridge 1976).

The second problem arose from an attempt to avoid the 5 h maintenance culture, very much along the same lines as was done successfully for the cytochemical section assay of LH (Chap. 11). It was argued that in the starved animal the fundus had been subjected to only very low concentrations of gastrin. Consequently, sections were made of the fundus from starved guinea-pigs and these indeed responded to exogenous gastrin. The drawback was that when sections were exposed to the carrier alone, they sometimes showed low carbonic anhydrase activity, as was expected, but occasionally they contained high levels of activity, exceeding the activity induced by low concentrations (5 and even 50 fg/ml) of exogenously added gastrin. Thus in this study, as in those on luteinizing hormone, there seemed to be a need to have the cells 'geared' or 'harnessed' to the hormone, even if the concentration of hormone used to harness the apparatus of the cell was below the limit of detection by the assay employed (as it is in the section assay for ACTH). Consequently, segments of gastric fundus were maintained at pH 7.6 (which gave best histological preservation of the cells) for 5 h and were then primed for 5 min (the time to achieve maximal effect in the segment assay) with G_{17}-gastrin at 2.3×10^{-16} mol/litre, namely at one-tenth of the least concentration detectable by this assay (Loveridge et al. 1978).

II. Assay

1. Procedure

Strips (segments) of the fundus of guinea-pigs (as used in the segment assay) are maintained individually for 5 h as in the segment assay but with the pH of the Trowell's T8 medium at pH 7.6. The culture medium is then replaced by fresh T8 medium containing 0.5 fg gastrin (G_{17}-gastrin) per millilitre of culture medium. After 5 min the strips are chilled to $-70°$ C in *n*-hexane, as discussed previously. They may be stored in dry glass tubes at $-70°$ C for up to 3 days but should not be used after this.

Sections are cut at 18 μm in a cryostat, as described in Chap. 5. The sections are placed in the compartmented trough (Fig. 12.3) so that duplicate sections can be treated with one of four logarithmically graded concentrations of the standard preparation of the hormone (G_{17}-gastrin), from 5 fg/ml to 5 pg/ml, or to one of two dilutions (usually 1 : 100 and 1 : 1000) of each plasma. The standard preparation of the hormone and the plasmas are diluted in 0.1 *M* HEPES buffer, pH 7.0 containing 0.005% gum tragacanth which helps to stabilize the histological appearance of the sections.

The plunger is inserted, so releasing the dilutions of the standard preparation of the hormone and of the plasma samples (e. g. 1 ml) on to the relevant sections. After 75 s the lid containing the syringes is removed and the sections are covered with the cytochemical reaction medium for demonstrating carbonic anhydrase activity (this chapter, Sect. B.I.2.). After treatment with H_2S, they are rinsed in distilled water and mounted in the water-soluble Farrants' medium. The reaction product is measured in the parietal cells, and 'blank' readings are made in the layer of muscle, as described in this chapter Sect. B.I.4.

2. Precision

Serial sections exposed to one concentration of G_{17}-gastrin for various periods of time showed that the maximal stimulation of carbonic anhydrase activity occurred at 75 s and was sufficiently extended to serve as the optimal time for this assay. Other sections subjected to the buffer and gum tragacanth (without hormone) showed low and stable activity over 2 min, the activity being considerably less than that induced by the hormone at a concentration of 5 fg/ml. When various concentrations of the standard preparation of gastrin were assayed for the activity induced in sections exposed to the hormone for 75 s, there was a linear, logarithmic dose-response from 5 fg to 5 pg/ml (2.3×10^{-15} to 2.3×10^{-12} *M*). The response to different concentrations (1 : 100 and 1 : 1000) of normal plasma was parallel to this dose-response graph and was abolished or virtually abolished either by treating the plasma with activated charcoal or by the addition of an antibody (at $1:10^4$ dilution) specific to gastrin.

The mean index of precision in eight assays was 0.10 ± 0.05; the fiducial limits of an assay (P, 0.95) were 95%–134%. In three assays, interassay variation was ± 16.3%; intra-assay variation was ± 6.4% (*n*, 4). The accuracy of the method was examined by the conventional recovery study: in two studies the recovery of gastrin added to plasma samples was 96% and 112% (Loveridge, Hoile, Johnson, Gardner and Chayen, unpublished results).

D. Validation

I. Effect of Cholecystokinin

The whole molecule of CCK has been shown to act rather like a competitive inhibitor of gastrin (Johnson and Grossman 1971). In the section assay CCK (95% pure; kindly supplied by Professor Mutt) showed activity, with the same time course, but it was 1000 times less potent than the heptadecapeptide gastrin. However, when various concentrations (0.05–500 pg/ml) of CCK were added to segments of gastric fundus, together with a single concentration (0.05 pg/ml) of G_{17}-gastrin, they markedly altered the effect of the gastrin (Fig. 12.4). Thus an equal weight of CCK decreased the gastrin-like activity to about a half of what it would have been had the 0.05 pg/ml of gastrin been present alone. When the CCK was present at 5 pg/ml, the depression in activity was only slight (16%), but when it was added at 50 pg/ml it increased the gastrin-like activity to twice that produced by 0.05 pg/ml of G_{17}-gastrin alone.

The dose response graph (Fig. 12.4) was not parallel to that of different concentrations of gastrin alone. Thus the presence of significant amounts of CCK (or some similar agent) would be indicated by this lack of parallelism. Although a sample which contained an equal weight of CCK would have assayed as having half the correct amount of G_{17}-gastrin, increasing amounts of CCK up to 5 pg/ml at least, would cause less discrepancy in the assay.

In a later study (Loveridge and Hoile, unpublished results) in which the cytochemical section assay was used, three samples were prepared, each containing 50 pg/ml (2.3×10^{-11} *M*) of G_{17}-gastrin. Cholecystokinin was added to the first at a concentration of 150 pg/ml; to the second at 50 pg/ml and to the third at 10 pg/ml

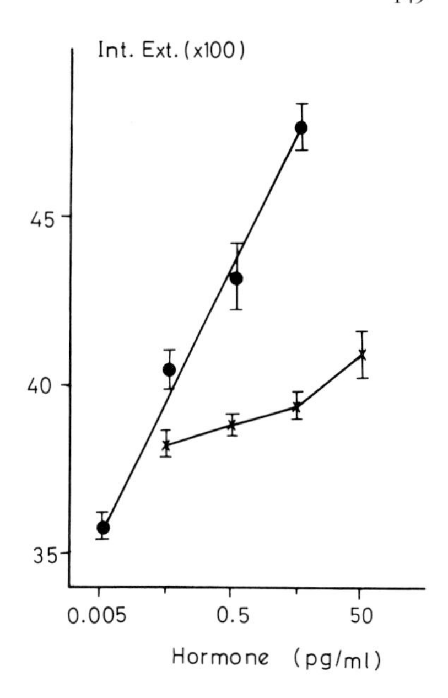

Fig. 12.4. The effect of various concentrations (0.05–50 pg/ml) of CCK on the carbonic anhydrase activity (Int. Ext. × 100) induced by a single concentration (0.05 pg/ml) of G_{17}-gastrin (*crosses*). The normal response to various concentrations of the G_{17}-gastrin alone (*filled circles*), in serial sections of the same tissue, is shown for comparison. Bars indicate the values measured in each of the duplicate sections

(3.75×10^{-11}, 1.25×10^{-11} and 2.5×10^{-12} *M* respectively). The samples were measured at 1:100 and 1:1000 dilutions against a standard dose-response to G_{17}-gastrin alone. The first two samples gave dose-responses which were not parallel to the standard calibration graph whereas the response to the two dilutions of the third sample was almost parallel, giving results close to those expected for that concentration of G_{17}-gastrin alone. Yet even though the dose-responses caused by the higher concentrations were not parallel to that of G_{17}-gastrin alone, at the 1:1000 dilution the samples 'assayed' at 38 and 40 pg/ml respectively. Thus at high dilutions the interference caused by CCK becomes only slight.

II. Effect of the Octapeptide of CCK

When tested alone, CCK-OP was equipotent on a molar basis with G_{17}-gastrin, as would be expected from the similarity of this peptide with the biologically active part of G_{17}-gastrin. However, as will be discussed in Chap. 15, when tested together with G_{17}-gastrin, it made no contribution to the activity induced by the gastrin alone, even when present at concentrations 100 times greater than the gastrin.

III. Effect of Secretin

When tested alone, secretin (95% pure, kindly supplied by Professor Mutt) was 100 times less potent than CCK (and therefore 100000 times less potent than G_{17}-gastrin) in stimulating carbonic anhydrase activity. This slight activity could have been due to contamination. However, when tested in the presence of G_{17}-gastrin it inhibited the carbonic anhydrase activity induced by the gastrin, the degree of inhibition being more marked as the concentration of secretin was increased over that of the gastrin. Thus 10^{-10} *M* secretin inhibited all the response that could be expected from the gastrin (10^{-13} *M*) alone. When secretin was present at 10^{-11} *M*, the inhibition was only 21%; at 10^{-12} *M* it was 16% and at 10^{-13} *M* (i.e. at equal molarity) secretin did not inhibit the action of G_{17}-gastrin. Consequently it seems that the concentration of secretin will need to be ten times that of the gastrin to cause any marked inhibition of gastrin-like activity.

IV. Effects of Other Mediators of Secretion

Acetylcholine and histamine are much implicated in the secretion of gastric acid. Although they are unlikely to occur in plasma to any appreciable extent, their effect on the activation of carbonic anhydrase in parietal cells in vitro has also been examined. It seems at present that although they may facilitate the response to gastrin, and so may alter the time course of the response, they do not influence the maximal stimulation produced by a given concentration of gastrin.

V. Effect of Big Gastrin

The G_{34}-gastrin (kindly supplied by Professor R. A. Gregory) had only a slight effect on the carbonic anhydrase activity of the parietal cells when tested under the conditions of this assay, namely at 5 min in the segment assay. This does not rule out the possibility that it may have a different time course of action (as indicated by Walsh et al. 1975) so that at some extended time it might have considerable biological activity. Similarly, Askew et al. (1979), using the cytochemical section assay, claimed to find no activation of carbonic anhydrase activity by G_{34}-gastrin, but they tested the activity at the time at which G_{17}-gastrin produced its maximal response. By analogy, if the long-acting thyroid stimulators (Chap. 10) had been tested only at the time at which TSH produces its maximal effect, it would have been concluded that these stimulators were not stimulatory.

VI. Trasylol: Measurement of Little and Big Gastrin Activity

Although G_{34}-gastrin gives little response in this assay at the time at which G_{17}-gastrin (and pentagastrin) produces its maximal effect, it is likely to contribute to the total gastrin-like activity in the plasma assessed over a longer time. This will occur either because its influence may take longer to be expressed, or because it will be altered, by tryptic activity in the plasma, to G_{17}-gastrin. Moreover, it has not yet been established how quickly the heptadecapeptide is degraded by such enzymes when it is stored, in plasma, even at low temperatures prior to being assayed. Thus it may be advantageous to divide each sample of plasma into two, one containing an inhibitor of trypsin-like enzymes (such as aprotinin (Trasylol), e. g. at 1000 units/ml), and to assay for gastrin-like activity after storing the plasma samples for a few days. Since trypsin can convert G_{34}-gastrin to G_{17}-gastrin (Yalow and Berson 1971), any G_{34}-gastrin will have been converted into the heptadecapeptide in the absence of the inhibitor, whereas the sample containing the inhibitor will give a value for G_{17}-gastrin alone, with little or no contribution from the bigger forms of this molecule.

VII. Comparison of the Cytochemical Bioassay and Radio-immunoassay

In the earliest study (Loveridge et al. 1974) there was good agreement between the bioassay and radio-immunoassay results on a sample from a patient with the

Zollinger-Ellison syndrome, and from two samples from one subject at different times after a meal. In two samples taken from a fasting subject, radio-immunoassay gave higher results than did the cytochemical segment assay. This discrepancy might be due to the presence of a greater concentration of big gastrin in the fasting state, but this was not investigated. The cytochemical bioassay gave activities equivalent to 34 and 13.3 pg/ml of pentagastrin activity in the fasting samples, and 118 and 90 pg/ml after food.

In spite of this apparent correlation between the cytochemical bioassay and radio-immunoassay, it would be surprising if such a correlation was found generally. This is because radio-immunoassay is normally done on serum: during the time taken to clot the blood and separate the serum, considerable degradation of the active form of gastrin must be expected, because of the tryptic activity present in plasma. In separate studies. Hoile (1979) sometimes found virtually no gastrin-like activity in the cytochemical bioassay in serum, even though there was normal activity in the plasma taken at the same time. The difficult interpretations forced on workers who use serum rather than plasma when investigating short-lived, protease-sensitive, polypeptide hormones, will be discussed later, in relation to gastrin-like activity induced by test meals (see below)

E. Physiological Studies

I. Effects of a Test Meal

Radio-immunoassay of gastrin is normally made on samples of serum. As discussed briefly above, this seems theoretically inadvisable, even though it may be technically necessary, in view of the fact that the proteolytic enzymes in the blood are free to act on the polypeptide hormone throughout the time taken to clot the blood and separate the serum. Presumably it may be argued that the antigenic determinants will remain during this process to be assayed by radio-immunoassay even if the biological activity of the hormone is impaired. The results obtained on the effects of test meals indicate that the full story may be considerably more complicated.

One of the findings that has led to doubts as to the involvement of gastrin in the secretion of gastric acid, despite the apparently clear evidence of various workers (particularly Gregory and Tracy 1961), is that acid secretion, induced by a test meal or by distension of the stomach, long precedes the rise in circulating levels of gastrin as measured by radio-immunoassay (e. g. Stagg et al. 1971).

Hoile and Loveridge (1977) began by establishing that the circulating levels (mean $\pm$SD) in 19 samples taken from 11 normal, fasting subjects were 12.5$\pm$6.1 pg/ml (relative to G_{17}-gastrin). They then found, in two fasting subjects given a test meal containing Oxo (2 cubes of Oxo in 150 ml of water: Byrnes et al. 1970; Stagg et al. 1971) to stimulate gastric acid secretion, that there was a rapid rise in the concentration of gastrin, assayed by the cytochemical section assay. In the first subject, two samples taken fasting gave values of 9.6 and 6.1 pg/ml (relative to G_{17}-gastrin); 5 min after the meal the concentration had risen to 180 pg/ml, declining to 5.6 pg/ml 20 min after the meal had been taken. In the second subject the results were 18.6 and 9.3 pg/ml fasting; 289 pg/ml 5 min after the meal, declining to 22.3 pg/ml 15 min later (Fig. 12.5).

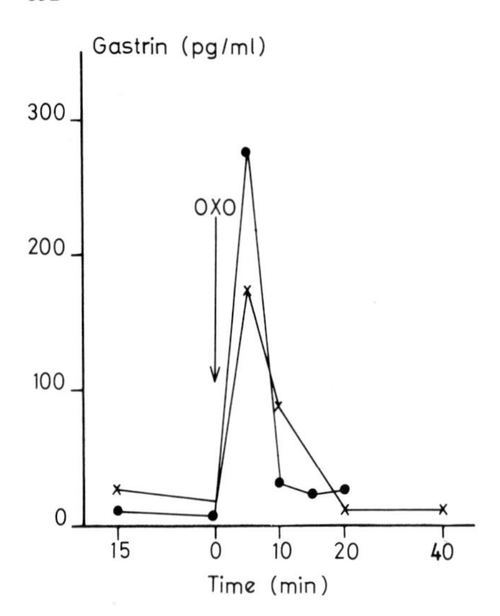

Fig. 12.5. The effect of an Oxo test meal on the circulating levels of gastrin-like activity (measured relative to G_{17}-gastrin) in two different subjects

This dramatic and rapid rise in the circulating level of bioreactive gastrin (measured in plasma) is in accord with its possible involvement in gastric acid secretion under physiological conditions. But the results are totally discordant with those apparently found by radio-immunoassay of serum. It will be remembered (as recorded above) that the cytochemical bioassay frequently records very little gastrin activity in serum.

To some degree this discrepancy seems to have been resolved by the results of Christofides et al. (1978) who found a similarly early rise, with a peak at 5 min, in circulating levels of gastrin, measured by radio-immunoassay of the *plasma*, when the stomach was distended by fluid. Although at first sight this seems a satisfactory resolution of the apparent discrepancy between radio-immunoassay and bioassay, we are left wondering what the radio-immunoassayists have been measuring that rises more than 20 min after the test meal. It may be of interest that this is the time when the circulating levels of CCK might be expected to rise, and that the cytochemical bioassay did not record its influence, at least at 20 min after the meal had been taken.

II. Studies on Pernicious Anaemia

The study by Bitensky et al. (1979; also Loveridge 1977) on the inhibition of gastrin responsiveness by antibodies directed to the parietal cell is of interest for a number of reasons and illustrates something of the versatility of the cytochemical bioassay system.

It began when four samples were assayed under code, two being known to be derived from normal subjects and two from patients with pernicious anaemia. Samples A and D, at 1 : 100 and 1 : 1000 dilution, gave dose-responses which were parallel to the standard dose-response graph (as in Fig. 12.2); the dose-response graphs of B and C were distinctly non-parallel and therefore the concentration of gastrin could not be assayed. It was therefore necessary to test whether there was some matter in B and C which interfered with the assay of any gastrin which might have been present.

Gastrin (G_{17}-gastrin) was added to an aliquot of each sample at a concentration of 500 pg/ml (so that at the first dilution, namely $1:10^2$, they should show 5 pg/ml). The samples were then tested at dilutions of up to $1:10^5$. Samples A and D again assayed normally, giving reasonable recovery of the added gastrin at all dilutions. Samples B and C gave non-parallel responses over the range $1:10^2$ to $1:10^4$ but the carbonic anhydrase activity stimulated by concentrations of $1:10^4$ and $1:10^5$ was parallel to that induced by suitable concentrations of the standard preparation of the hormone (Fig. 12.6). Indeed, calculated from the parallel parts of these graphs, and subtracting the concentration of exogenously added gastrin, it appeared that sample B contained gastrin at a concentration of 100 pg/ml and sample C at 250 pg/ml, although it is difficult to be precise about these concentrations because of the considerable dilution at which the samples were assayed.

As a further test, the samples were first diluted at $1:10^2$ to $1:10^4$ and G_{17}-gastrin was added to each diluted sample to achieve a concentration of 5 pg/ml in each. Samples A and D gave the correct assay at all dilutions, namely 5 pg/ml plus the

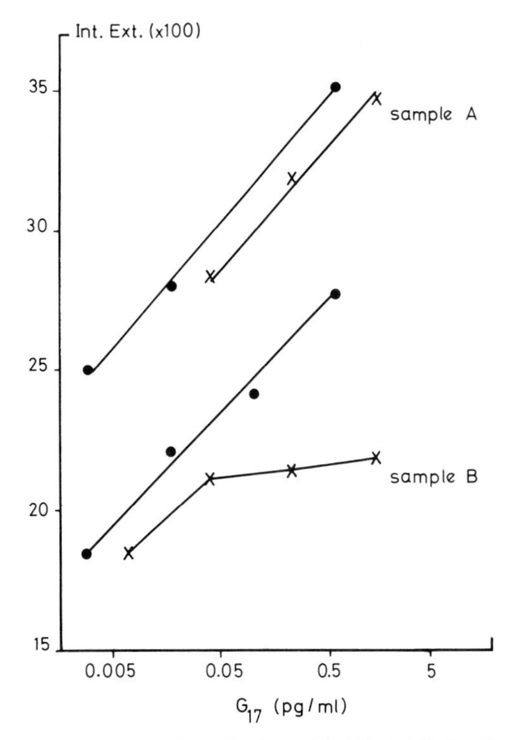

Fig. 12.6. The carbonic anhydrase activity (Int. Ext. × 100) induced by dilutions of $1:10^2$, $1:10^3$, $1:10^4$ and (sample B) $1:10^5$ of samples A and B to which exogenous G_{17}-gastrin had been added to achieve a concentration, at the least dilution ($1:10^2$) of 5 pg/ml. The response of serial sections, from the same tissue, to graded concentrations of a standard preparation of G_{17}-gastrin is shown (*filled circles*) for comparison. The activities (*crosses*) induced by sample A are parallel to those induced by the standard; those induced by sample B are distinctly not parallel to the standard response except at dilutions of $1:10^4$ and $1:10^5$

expected concentration of the endogenous gastrin. The other samples gave activities far below 5 pg/ml except when the sample had been diluted $1:10^4$.

Thus there appeared to be some material in samples B and C which interfered with the ability of the endogenous gastrin, or of exogenously added G_{17}-gastrin, to influence the carbonic anhydrase activity of the parietal cells (in the section assay). Moreover, when this material was sufficiently diluted, its influence became insignificant. There seemed to be three obvious possibilities: (1) there was some material which interfered directly with carbonic anhydrase; (2) there was something, like an antibody to gastrin, which interefered with the expression of the hormone; or (3) there was some matter, such as an antibody to the receptor, which inhibited the effect of the hormone on the cell membrane.

To test these influences, segments of guinea-pig fundus were maintained in vitro for 5 h in the normal way (at pH 7.0) except that the medium contained one of the samples diluted 1:100 in the T8 medium. As controls, others were maintained with the T8 medium alone. At the end of this period the medium was replaced by fresh T8 medium containing gastrin at either 5 or 0.005 pg/ml, without the sample. After the appropriate time the segments were chilled, sectioned and tested for carbonic anhydrase activity. The segments maintained in the presence of samples A and D showed the same response to the two concentrations of gastrin as did those maintained in the presence of T8 medium alone. Thus these samples had no effect on the ability of the parietal cells to respond to gastrin. In contrast, the segments which had been maintained for 5 h in the presence of sample C showed virtually no stimulation by gastrin even though when the hormone was added there was none of this sample present, unless bound to the tissue. (There was not enough of sample B to use in this test.)

Thus the evidence suggested that sample C (and the same may apply to sample B) contained something, probably an antibody, which reacted with some component of the cell, possibly the gastrin receptor, to inhibit the ability of the cell or cell surface to respond to gastrin. At high dilutions this material became 'diluted-out' and allowed the expression of the hormone. Later it was found that samples B and C came from the patients with pernicious anaemia.

This observation was then extended. Antibodies directed either to the parietal cell or to intrinsic factor were obtained by precipitation of immunoglobulins by 40% ammonium sulphate from appropriate sera containing one or other type of antibody. Antibodies to the parietal cells were verified by immunofluorescence on human stomach. Antibodies to intrinsic factor were determined by the charcoal-binding radio-immunoassay. The antibodies were added at $1:10^2$ dilution to the T8 medium used for the 5-h maintenance of segments of guinea-pig fundus. At the end of this period the medium was replaced by fresh T8 medium containing gastrin at either 5 pg/ml or 5 fg/ml. Those segments which had been maintained in the presence of intrinsic factor antibodies responded normally to the gastrin but there was no response in the segments which had been maintained in the presence of the parietal cell antibodies. Thus the latter behaved as did the segments maintained in the presence of a dilution of sample C.

Thus these studies suggest that the achlorhydria of pernicious anaemia might be caused by the interference of the response of the parietal cells to gastrin caused by the presence of parietal cell antibodies. Furthermore, such antibodies would be expected to interfere with the trophic effect of the hormone and this may in part account for the atrophic gastritis associated with this condition. The trophic effects of trophic hormones have received unduly little attention.

This study (Bitensky et al. 1979) has been discussed in some detail because it shows how a non-parallel response can be informative. If the response becomes parallel at very high dilutions, and this can be examined because of the remarkable sensitivity of the cytochemical bioassays, it is possible to analyse the interference. This analysis may require recourse to a segment assay system, as in this case. And occasionally the analysis is more rewarding than a simple assay would have been.

Chapter 13

Parathyroid Hormone

A. Background Information

I. The Hormone

There is a vast amount of information concerning PTH, its role in the maintenance of calcium homeostasis, and its possible implications in diseases of bone. Its function is intimately linked with vitamin D and its metabolites and possibly with calcitonin. Furthermore, the many different actions ascribed to PTH may have very different dose-response relationships. It is beyond the remit of this monograph to review this literature: reference should be made to the articles by Parsons and Potts (1972); Parsons et al. (1975); Parsons (1976).

Bovine, porcine and human PTH consist of 84 amino acids. All contain one methionine residue at position 8; the bovine hormone has a second methionine residue at position 18 (numbering from the amino-terminus). Oxidation of the methionine residue, or residues, results in considerable loss of activity (as discussed later). The first amino acid of the porcine and human hormone is serine instead of alanine as in the bovine PTH. This first amino acid can have a powerful influence on the activity of the molecule. Thus, for example, the hormone is readily cleaved in vivo to a 1–34 fragment which has 80% of the activity of the intact 1–84 molecule in the rat kidney adenyl cyclase assay, and 130% of the activity of the whole molecule when assayed by the chick hypercalcaemia method. Yet the 2–34 fragment has perhaps 2% of the activity of the 1–84 molecule in the former, and 65% in the latter assay. By neither method was the 3–34 fragment found to be active (Parsons et al. 1975). Of even greater moment is the finding that the desamino^1-bPTH$_{1-34}$, which lacks only the amino group of the first amino acid, had undetectable activity in the rat kidney adenyl cyclase assay, while retaining 55% of the activity of the whole 1–84 molecule in the chick hypercalcaemia assay (Parsons et al. 1975). This is a remarkable case of dissociation between adenyl cyclase activity and the normal physiological effect of a hormone.

The hormone appears to be liberated as the 1–84 molecule but rapidly suffers cleavage (Segre et al. 1974); the kidney and liver are capable of such cleavage (Fujita et al. 1977). The half-time in the circulation is about 6 min (as reviewed by Parsons et al. 1975). The N-terminal fragment, which will still have considerable bioactivity provided that it retains the first 34 amino acids, is also believed to have a very short half-time in the circulation whereas the larger C-terminal fragment may persist for up to 20 times as long (Segre et al. 1972). This prolonged retention of the C-terminal fragment in the circulation possibly accounts for the ability of radio-immunoassays, with antibodies directed to this end of the molecule, being able to detect abnormally high concentrations of the hormone in hyperparathyroid patients (Arnaud et al. 1974).

The circulating level of the biologically active hormone, either the 1–84 or the 1–34 molecule, has been calculated from the rate at which PTH had to be infused into parathyroidectomized dogs in order to maintain the normal level of plasma calcium. Taking into account the half-time of the biologically active hormone, Parsons and Reit (1974) estimated that the normal circulating level of parathyroid hormone would have to be only about 10 pg/ml. In their own studies on unrestrained normal dogs they found that a constant infusion rate of 100 ng/kg/h caused marked hypercalcaemia; this too indicated that the normal circulating level could not be greater than 10 pg/ml. Other evidence, based on measurements made on the effluent blood from the external parathyroid glands of calves (reviewed by Parsons et al. 1975) in which the endogenous secretion rate was 0.5 ng/kg/min, corroborated the view that the circulating level of PTH was about 10 pg/ml.

In view of these findings and calculations, it may seem surprising that, until recently, radio-immunoassay measured values of up to 500 pg/ml in normal human subjects, with higher values measured in conditions associated with hyperparathyroidism. At least to some extent, this was due to the assays using antibodies directed to the carboxy-terminus and therefore measuring the long-lived fragment, which can accumulate in the circulation. The limit of sensitivity of such assays appears to be about 250–500 pg/ml. Even the more recent radio-immunoassay using an antibody directed to the N-terminus (Desplan et al. 1978) has a sensitivity of perhaps 75 pg/ml. Consequently none of these assays could measure the concentration of biologically active hormone if that level were indeed to be about 10 pg/ml. Moreover the European PTH Study Group (1978) found very wide disagreement between co-operating laboratories that were given identical samples to assay (by the C-terminal radio-immunoassay). For example, the assays on one sample varied from 0.097 ng/ml to 3.5 ng/ml; on another they ranged from 0.319 to 35 ng/ml. The ratio of the concentration of PTH in a sample diluted ten times, to the concentration in the original sample (which should have been 1 : 10) varied in the different laboratories from 1 : 2.9 to 1 : 13.5. It seemed, therefore, that a precise and sensitive cytochemical bioassay of PTH could meet a practical need.

II. Bioassays

It is amusing to reflect that the 8 kg of bone in an average adult represents one of the largest masses of target tissue of a hormone, yet that hormone is produced by the smallest endocrine glands, the parathyroids, the total weight of which may be 20–40 mg (Parsons and Potts 1972). As reviewed by Zanelli and Parsons (1979) most bioassays depend on the alterations in calcium induced by a particular range of concentrations of PTH; in general this must be induced by the hormone mobilizing calcium from bone. (However, it must be emphasized that PTH affects other components of bone as well, as discussed by Parsons and Potts 1972, and that Parsons and Reit 1974, could not detect any significant mobilization of bone calcium during minimum hypercalcaemic infusions in dogs.)

One of the most sensitive bioassays is the intravenous chick assay of Parsons et al. (1973). The PTH, in the presence of calcium chloride (20 μmoles/bird) is injected intravenously into chicks (7–14 days old). Sixty minutes later the chicks are anaesthetized and bled; the concentration of plasma-calcium is then measured. These 3+3 assays use 30 chicks in groups of five. The mean index of precision was 0.14. Although this is one of the most sensitive bioassays, it requires that the level of the hormone in the chick blood should be of the order of 10^{-8} g/ml (Parsons et al. 1975).

III. In Vitro Assays

In 1968 Chase and Aurbach (see Parsons et al 1975) showed that PTH would activate the adenylate cyclase of membranes isolated from broken rat kidney cells. Although with various improvements this system has been used as an assay of PTH, its sensitivity is still nearly five orders of magnitude greater than the expected circulating level of this hormone (Parsons et al. 1975). Adenylate cyclase is also activated by PTH in cells, isolated from rat or mouse calvaria, grown to confluent monolayers. The limit of sensitivity seems to be about 1 ng/ml although there have been occasional reports of remarkably greater sensitivity (as reviewed by Parsons et al. 1975).

B. Cytochemical Bioassay of Parathyroid Hormone

I. Background

1. Possible Physiological Functions of the Hormone

The ability of PTH to remove calcium from bone may require high concentrations of the hormone and may be strongly influenced by other factors, including the concentration of calcitonin (as discussed by Parsons and Potts 1972; Parsons et al. 1975). Thus at first sight bone did not seem a very promising target tissue although McPartlin et al. (1977) have shown, cytochemically, dose-related effects of PTH on the alkaline phosphatase activity of calvaria.

Parathyroid hormone has a sodium-linked phosphaturic effect (Agus et al. 1971) but this seems to be less sensitive than the calcium-retaining effect in the kidney (Parsons et al. 1975). By a micropuncture technique Agus et al. (1973) showed that this retention of calcium, induced by PTH, depended on increased calcium reabsorption by the distal tubule. Parsons et al. (1975) considered this effect to be among the most sensitive responses recorded for the action of PTH in vivo.

Although PTH increases the absorption of calcium by the intestine, Parsons et al. (1975) considered that this might be an indirect effect of increased renal 1-hydroxylation of 25-hydroxycholecalciferol, which is another function of PTH which interested us. Thus at the outset of these studies it seemed to us that the kidney cortex was the most promising target tissue.

II. Cytochemical Studies on the Responses of the Renal Cortex to PTH

1. General Investigations

In these studies (Chambers et al. 1978 b) segments of renal cortex of guinea-pigs (Hartley strain; about 450 g weight) which had been killed by asphyxiation in nitrogen, were examined either at death or after maintenance at 37° C in vitro for 5 h (as for the other cytochemical bioassays) both directly and after various periods of exposure (1–20 min) to purified bovine PTH (2500 U/mg; NIBSC Code 72/286), usually at a concentration of 1 pg/ml (approximately $10^{-13}M$). The results showed that it was possible to retain the normal histology of the renal cortex during this period of maintenance culture in Trowell's T8 medium, at pH values of about 7.6 and that the addition to the medium of 1m*M* choline occasionally improved the histological appearance of the tissue. It may be remarked that choline is a normal component of Eagle's minimum essential medium, which has been used by Fenton et al. (1978) in a later modification of the cytochemical bioassay of PTH.

Apart from retaining the histological structure of the tissue, the maintenance in vitro had little effect on the activity of certain enzymes. However, it markedly depressed the activities of glucose-6-phosphate dehydrogenase and of the diaphorase associated with it, and of alkaline phosphatase and carbonic anhydrase. When the general biochemical activity of a tissue is retained during in vitro maintenance, but certain activities are markedly depressed, it is tempting to test whether such depression is caused by the lack of one or other hormone. It will be remembered that one of the objects of this period of maintenance in vitro is to deprive the target cells of their specific hormone. Consequently attention was focussed on those enzymatic activities which were diminished at the end of the maintenance period.

2. Method of Relating Activity to Histology

Each region of the nephron is characterized by its three-dimensional position and by well-defined histological features such as the shape and size of the tubules. In sections of the guinea-pig renal cortex there is one distal convoluted tubule to about 12–20 proximal convoluted tubules. The former can be distinguished by its shape (and the shape of the cells); by its lack of alkaline phosphatase; and by its position (because it lies close to the glomerulus and its related arteriole (Fig. 13.1). The proximal convoluted tubules and the straight portion of the proximal tubule, the pars recta, are rich in alkaline phosphatase; the former has a well-marked brush border. It was demonstrated to us by Dr. Schäfer that when sections were reacted for glucose-6-phosphate dehydrogenase activity, particularly if nitro-blue tetrazolium was used as the hydrogen acceptor, they could then be reacted for alkaline phosphatase activity (by the Gomori calcium method). Then tubules which had the black deposit, denoting alkaline phosphatase activity, in their lumenal borders were obviously proximal tubules. This was valuable in mapping the enzymatic activities in the tubules as seen in sections, where otherwise their relationship to other tubules may be none too easy to define. In other studies sections were reacted for the enzyme to be investigated and then serial sections were reacted for alkaline phosphatase activity. Comparison of the two serial sections allowed us to define the location of the enzyme investigated.

3. Effect of PTH on Different Regions of the Nephron

Several enzymes, in different types of tubule, showed diminished activity after the 5-h maintenance period and a resurgence of activity after various times of exposure to the hormone. Only a few of these were investigated in the studies of Chambers et al. (1978 b).

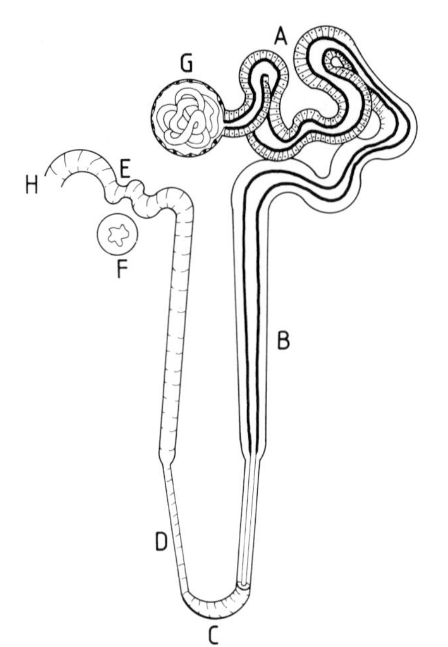

Fig. 13.1. Diagrammatic representation of a kidney nephron. *A*, proximal convoluted tubule, and *B*: straight portion of the proximal tubule (pars recta), both with alkaline phosphatase activity (*black border*). The loop of Henle (*C*) has a thinner descending tubule and thicker ascending tubule (*D*) leading to the straight part of the distal tubule and the distal convoluted tubule (*E*), which opens to the collecting tubule (*H*). The distal convoluted tubule (E) lies between the glomerulus (*G*) and the arteriole (*F*)

The characteristic alkaline phosphatase activity in the brush border of the cells of the proximal convoluted tubules decreased during the in vitro maintenance period but was stimulated up to 50% above its diminished value by PTH, at 0.1 pg/ml, acting for 4 min. The activity then declined so that, 4 min later, it was back to the basal level of the 5-h maintenance period. The response appeared to be even more rapid in female guinea-pigs.

Carbonic anhydrase activity was also present in the proximal tubules. It decreased as a result of the in vitro maintenance and was stimulated by PTH. In tissue from female guinea-pigs the stimulation reached a peak 4 min after adding the hormone. But when the segments, which had been maintained for 5 h, were given fresh medium for 8 min and then exposed to fresh medium containing the hormone, the activation was maximal after only 1 min, and was dose-related over the concentrations of PTH of 1 fg – 1 pg/ml.

The pars recta, especially at the juxta-medullary boundary, showed remarkably high activity of the NADPH diaphorase, i.e. it was strongly coloured when the sections were reacted for glucose-6-phosphate dehydrogenase activity without the use of an intermediate hydrogen acceptor. Under such conditions (see Chap. 5 for fuller details) the NADPH (generated by glucose-6-phosphate dehydrogenase acting on glucose-6-phosphate and $NADP^+$) must be reoxidized by the diaphorase to react finally with neotetrazolium chloride. This activity declined sharply during the 5-h maintenance culture (Fig. 13.2) and was reactivated, to the full degree found in the sample at death, by PTH at concentrations up to 1 pg/ml. This response was relatively slow, becoming maximal only after 16 min treatment; it increased linearly with increasing concentrations of the hormone from 1 fg to1 pg/ml (Chambers et al. 1976). This diaphorase activity is probably associated with the microsomal respiratory pathway (Chayen et al. 1974 a) and it is tempting to speculate that the high activity of this pathway in this region of the nephron might be related to the

NADPH requirement of the renal system that hydroxylates 25-hydroxycholecalciferol to the 1,25-dihydroxy derivative. There seems little doubt that this system in the pars recta merits further study.

Glucose-6-phosphate dehydrogenase activity (measured in the presence of an intermediate hydrogen acceptor) was present in most tubules of the cortex but most markedly in the distal convoluted tubules. It, too, declined during the 5-h culture and was restimulated by PTH. Although the hormone stimulated activity in both the distal and proximal convoluted tubules (and in others), the steepest dose-response was in the distal convoluted tubules. There was a linear relationship between the activity and the logarithmic concentration of the hormone applied over the range 0.1 fg/ml to 1 pg/ml; tissue exposed to 0.01 fg/ml showed the same level of activity as the control, which had received fresh medium lacking hormone (Chambers et al. 1976). In the simple culture system, as already described, the activity of the enzyme in distal convoluted tubules exposed for 16–18 min to the hormone was 2.5 times the basal level found at the end of the 5-h maintenance period. It was therefore decided to use this activity as the basis of a bioassay of PTH.

III. Cytochemical Bioassay

1. Studies Required to Produce an Assay

As just discussed, the activation of glucose-6-phosphate dehydrogenase in the distal convoluted tubules was the most strongly dose-dependent and marked activity that

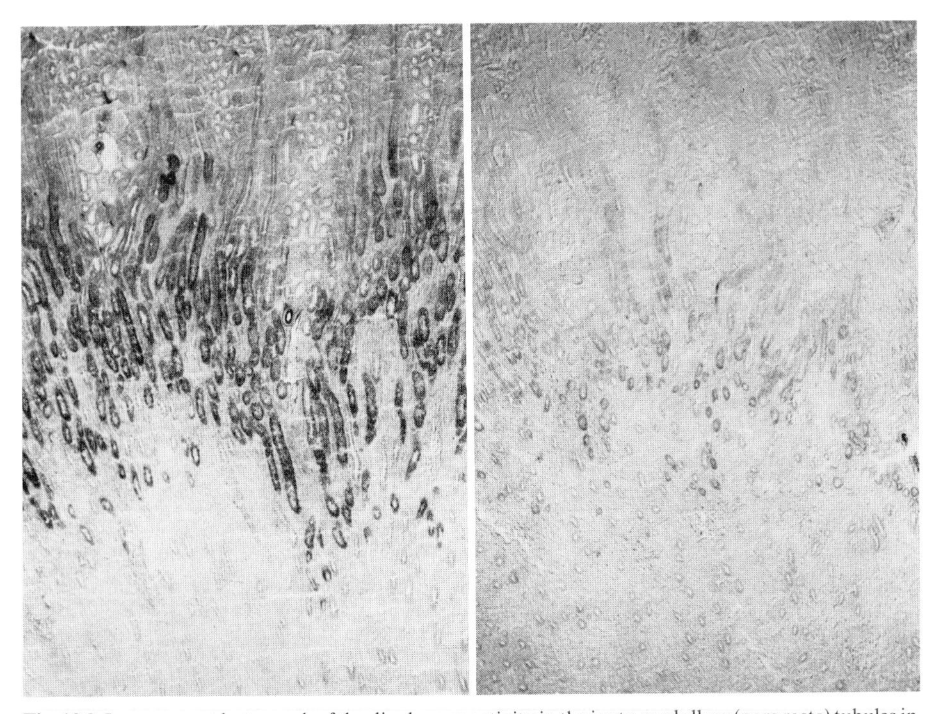

Fig. 13.2. Low-power photograph of the diaphorase activity in the juxta-medullary (pars recta) tubules in the biopsy sample (*left*) and after 5-h maintenance culture (*right*). The activity reverted to that shown on the left in other segments of the same kidney exposed to PTH (up to 1 pg/ml)

had been observed by us. However, a number of factors had to be investigated and clarified before it could form the basis of a bioassay.

First, the activity varied during the course of the year, becoming weak and responding to PTH poorly in late winter and early spring. However, the response could be maintained if the guinea-pigs were fed with cabbage and with a vitamin supplement in their drinking water. We have used two such supplements. Abidec and Adexolin (Glaxo). The latter contains 5,250 units of vitamin A, 105 mg of vitamin C and 35 μg of vitamin D per ml; 0.3 ml is added to every 200 ml of the drinking water (Chambers et al. 1978 a). The Abidec was given (Fenton et al. 1978) so that each 500 ml of drinking water contained 6000 units of vitamin A, 600 units of vitamin D, 1.5 μg of vitamin B_1, 0.6 μg of vitamin B_2, 0.75 μg of vitamin B_6, 7.5 μg of nicotinamide and of ascorbic acid. Initially these supplements were tried because it was thought that the animals might be deficient in vitamin D at these times of the year but the improvement in the response to PTH could be due to any of the components common to these supplements. This question requires further study.

Secondly, it was found that the cytochemical reaction-product for glucose-6-phosphate dehydrogenase activity was 'blotchy'. Such peculiar interference in the reaction could have been caused by metal ions accumulating in different regions of the cortex, and this phenomenon merits further examination. However, the interference was eliminated by adding potassium cyanide to the reaction medium. (Cyanide is a useful chelator of heavy metals including copper, which is often a contaminant of glycyl glycine used as a buffering agent.)

Thirdly, the responses produced by dilutions of plasma frequently were not parallel to those induced by a standard preparation of the hormone, diluted in T8 medium. This was simply overcome by including the same concentration of a PTH-deficient plasma in the solution containing the standard preparation as occurred in the dilutions of the plasma to be assayed. Moreover, the standard preparation of the hormone lost biological activity in this assay very rapidly when it was dissolved only in T8 medium even at $-70°$ C, and this degradation was markedly retarded when it was dissolved in plasma that was relatively free of PTH.

2. Assay Procedure (Chambers et al. 1978 a)

Female guinea-pigs (Hartley strain, 400–550 g) are fed for at least 1 week with rabbit pellets (BOCM/Silcocks Coney brand No. 350 was used but this is unlikely to be critical); cabbage is given daily and the drinking water contains either 0.3 ml of Adexolin vitamin drops (Glaxo; see above) in every 200 ml, or a suitable concentration of the Abidec preparation.

The guinea-pigs are killed by asphyxiation in nitrogen. Both kidneys are removed, decapsulated and halved sagittally. The medulla is excluded and the remaining cortex is cut into several segments of about 5 mm thickness and not exceeding 8 mm in their larger planar dimension. They are maintained separately for 5 h at 37° C with Trowell's T8 medium, pH 7.6, in an atmosphere containing 95% oxygen and 5% CO_2.

At the end of this 5-h period the medium is replaced by fresh T8 medium for 8 min to remove any disadvantageous metabolites which may have been secreted from the segments during the 5-h maintenance and which tend to keep the glucose-6-phosphate dehydrogenase activity somewhat high. The medium of each of four of the segments is then replaced by T8 medium, pH 7.6, containing one of a graded series of concentrations (1 fg/ml to 1 pg/ml) of the standard preparation of the hormone. The standard preparation of the hormone is diluted so that the highest concentration (1 pg/ml) contains 1% of a PTH-deficient plasma (e. g, from a patient after radical parathyroidectomy); the next lower concentration (0.1 pg/ml) therefore contains 0.1% of the plasma, etc., corresponding to the 1 : 100 and 1 : 1000 dilutions of the test plasmas. Each of two other segments is exposed to fresh medium containing one of two dilutions of the plasma in T8 medium, usually at concentrations of $1:10^2$ and $1:10^3$.

Tissue from animals fed with the vitamin supplement respond rather differently from tissue from animals not fed with this supplement. Thus the maximal response is now found at 8 min. This variability of the time of response means that it may be worthwhile to test the speed of response in each strain of guinea-pig under each set of conditions that is used. Under the conditions just described, the maximal response has been at 8 min. Consequently the hormone is allowed to act for this time and then the segments are chilled to −70° C and stored at this temperature for a maximum of 3 days. They are then sectioned at 16 μm (to ensure linearity of the enzymatic reaction with thickness of the sections and with time of reaction) and the sections are reacted to disclose glucose-6-phosphate dehydrogenase activity.

The reaction medium is composed of the following: glucose-6-phosphate, 5 m*M*; $NADP^+$, 3m*M*; phenazine methosulphate (the intermediate hydrogen acceptor), 0.67 m*M*; neotetrazolium chloride, 5m*M*; potassium cyanide, 10 m*M*; polyvinyl alcohol (12% G18/140 Polyviol, from Wacker-Chemie GMBH, Postfach 1260, 8263 Burghausen/Obb., West Germany, or preferably 30% of GO4/140 Polyviol from the same source); all in 0.05 *M* glycyl glycine buffer at pH 8.0. The reaction is done at 37° C in an atmosphere of nitrogen (Chap. 5) and is usually sufficiently strong for adequate measurement after 3–5 min.

The sections are washed in distilled water and left to dry. They may be stored, dry, in the dark until they are to be measured. They are then mounted in the water-miscible Farrants' medium and the medium is left to settle for a few minutes. The amount of formazan (reaction product) in distal convoluted tubules is measured with a Vickers M85 scanning and integrating microdensitometer with a ×40 objective; a mask of 8 μm diameter or larger (up to the breadth of the cells to be measured) and the smallest size of scanning spot; the wavelength is 585 nm, which is the isobestic point of the two formazans of neotetrazolium chloride (Butcher and Altman 1973; also Chap 6).

IV. Validation

Human plasma, whether from normal subjects or from patients with hyper- or hypoparathyroidism, gave linear log-dose-responses which were parallel to those obtained with purified 1–84 bPTH (Fig. 13.3).

In one study, part of one sample of bPTH was treated with hydrogen peroxide to oxidize the methionine residues and so inactivate the hormone (Tashjian et al. 1964). Only 10% of the activity remained, when tested in the chick hypercalcaemia in vivo bioassay of Parsons et al. (1973). The cytochemical bioassay also showed 90% inhibition of the initial activity (Chambers et al. 1978 a). The activity in normal plasma was inhibited by at least 90% when the plasma was treated with a specific antibody to PTH for 5 min (Fig. 13.3). The antibody (Burroughs Wellcome 211/32)

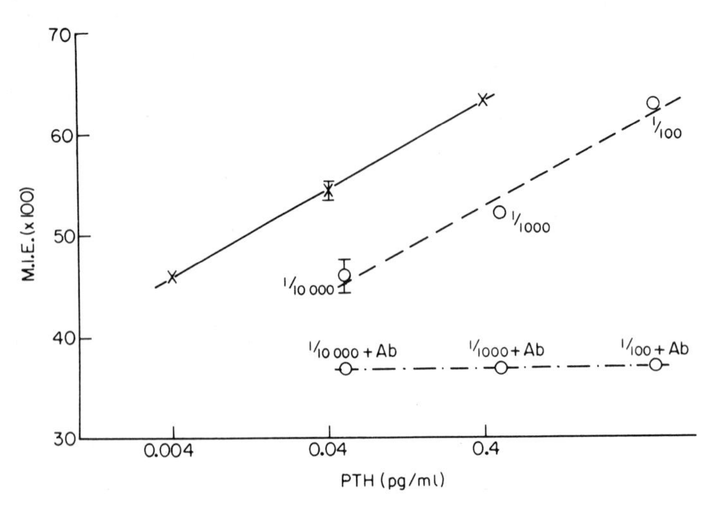

Fig. 13.3. The response produced by dilutions of a human plasma (*broken line*) in the glucose-6-phosphate dehydrogenase activity (M.I.E. × 100) in distal convoluted tubules in segments of guinea-pig kidney cortex. It is parallel to that elicited by graded concentrations of a standard preparation of bovine parathyroid hormone (*solid line*), and was annulled by the addition of the antibody (Ab). Bars show the mean activity in duplicate sections; where no bars are shown, the values were identical. (Chambers et al. 1978 a)

ed with sites at both the amino- and carboxy-terminus and was used at a final dilution of 1/20,000 before the sample was diluted $1:10^2$ and $1:10^3$.

The limit of sensitivity of the assay was 12.5×10^{-9} IU/ml (5 fg/ml, bPTH); the response was linear over the range 12.5×10^{-9} to 12.5×10^{-6} IU/ml (or 5 fg/ml to 5 pg/ml). The mean index of precision (±SEM) was 0.12 ± 0.07 (*n*, 23) and the fiducial limits (P, 0.95) of individual assays have been within 71%–140%. Interassay variation has been within ±13% (Chambers 1979; also Chambers et al. 1978 a). A known amount of PTH, added to a plasma, was 87% recovered by the assay (Chambers 1979). By the use of a very similar procedure, Fenton et al. (1978) found a mean index of precision (±SEM) of 0.13 ± 0.012 (*n*, 19) with fiducial limits (P, 0.95) of 78%–131%. The limit of sensitivity was 0.1 fg/ml. Arginine vasopressin, at between 1 and 100 pg/ml, and 1,25-dihydroxycholecalciferol (10–10,000 pg/ml) did not significantly influence the activity of glucose-6-phosphate dehydrogenase in segments used in this assay system; although high concentrations of calcitonin slightly influenced this enzymatic activity, the relative molar potency of calcitonin was 1×10^{-7} compared with that of PTH. Chambers (1979) found a similar effect of high concentrations of calcitonin, but no effect by concentrations of 20–200 pg/ml. Pitressin had no effect on this assay system.

The normal range of the circulating concentration of the hormone was reported as 1.0 to 15.5 pg/ml, based on human PTH, by Chambers et al. (1978 a), with samples from hypoparathyroid patients containing 0.13 and 0.15 pg/ml, and with 100 pg/ml from a hyperparathyroid patient. Fenton et al. (1978) found normal samples to contain between 1 and 30 pg/ml, based on the bovine standard; three hypoparathyroid plasmas assayed at 0.7, 0.4 and 0.9 pg/ml, whereas eight hyperparathyroid plasmas gave values ranging from 34 to 311 pg/ml, with a mean of 109 pg/ml.

As with all the cytochemical bioassays, possibly the best validation of the assay comes from physiological studies. It is well known that the circulating levels of PTH are closely related to those of ionized calcium; this is the basis of the hypercalcaemia bioassay in the chick (Parsons et al. 1973). Consequently a slow infusion of calcium, as calcium gluceptate, was given to a volunteer at the rate of 2 mg/kg/h over a period of 3 h, and the PTH levels were measured by the cytochemical bioassay. There was a rapid drop in the circulating level of the hormone shortly after the infusion began (Fig. 13.4), from 2.5 to 0.36 pg/ml. Similarly in a volunteer given 10 mg of EDTA/kg over a period of 2 h, the circulating level of PTH measured by the cytochemical bioassay rose from 3 pg/ml to 9.5 pg/ml 15 min after the infusion began, reaching 13 pg/ml 30 min after the onset of the perfusion (Chambers 1979). The changes in PTH levels correlated well with changes in ionized calcium in the circulation. Thus the activity measured by the cytochemical bioassay behaves as is expected of PTH.

Even at the time of writing, this assay is well established in three laboratories in Great Britain and one in Canada, with two other centres having workers who are now trained in the assay. Because of the great differences between what is measured by radio-immunoassay and the expected levels of biologically active hormone, which are actually measured by the cytochemical bioassay, it is likely that this cytochemical bioassay may be of even greater importance than the previous assays. The assay also opens a great amount of new investigatory possibilities, including studies on the relative functions of the 1–84 molecule and its fragments; and the role of PTH in conditions affecting bone, such as osteoporosis of the elderly, and in rheumatic diseases generally.

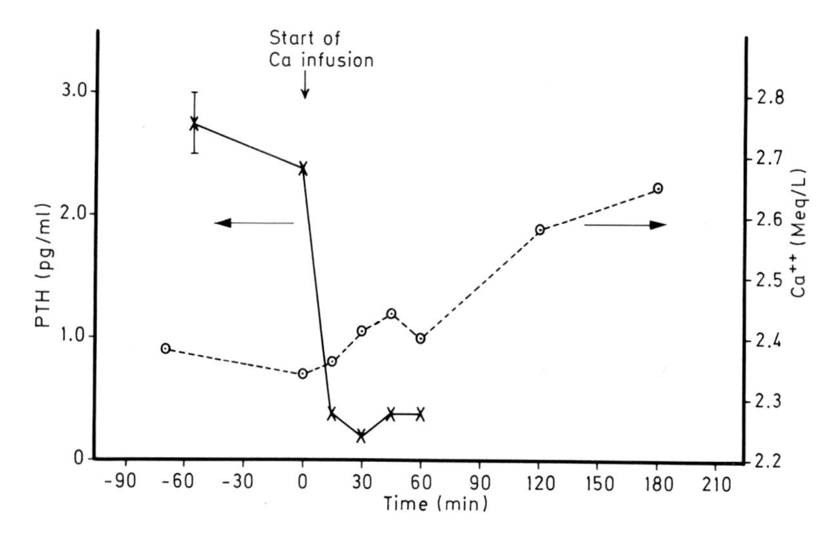

Fig. 13.4. The effect on the circulating levels of PTH-like activity, as measured in the cytochemical bioassay, of an infusion of calcium into a human subject. The circulating level (*crosses*) dropped rapidly after the infusion of calcium began, changing from 2.5 to 0.36 pg/ml, and stayed at this level for the next 45 min. The circulating levels of ionized calcium (Meq/L) are shown by the *open circles* and the *broken line*

Chapter 14

Hypophysiotrophic Hormones: Two-Step Bioassays

A. Background

I. Hypothalamus as an Endocrine Organ

The endocrinological importance of the pituitary gland has long been recognized but it is only more recently that it has become appreciated that pituitary activity (or activity of the hypophysis) can be controlled by small peptides released from the hypothalamus. Moreover, many regions of the brain influence the release of such hypothalamic peptides so that the endocrine system generally may be, at least in part, under the influence of phenomena previously considered to be more the province of psychology than endocrinology. Thus the hypothalamic hormones, synthesized in as yet undefined cells of the hypothalamus (Buckingham 1978), are secreted down axons and stored in the terminal dendrites in the region of the median eminence (Besser 1974 a). The release of the stored hormones, which are present in nanogram quantities in the median eminence (Besser 1974 a), appears to be under complex neurological control including time-dependent rhythms, psychological factors such as stress and possibly 'mood', as well as control by feed-back mechanisms and reflex responses mediated by peripheral nerves (Besser 1974 a; Buckingham 1978). The release of hypothalamic hormones to the pituitary gland can also cause transport of these hypothalamic hormones into the cerebrospinal fluid (Buckingham 1977, 1978), so accounting for their presence in this fluid. Some of these hypothalamic hormones cause the release of polypeptide hormones from the pituitary gland; others inhibit this release (Besser 1977).

It used to be said that the pituitary gland was the conductor of the endocrine orchestra. It now seems that its function is more analogous to that of the leader of the orchestra, with the hypothalamus as the conductor. Relatively minute amounts of stimulating or inhibiting peptides from the hypothalamus produce a considerable response from the pituitary gland (the circulating levels of pituitary hormones being measured in picogrammes per millilitre of blood) and this pituitary response then produces a 'cascade' of response from the target organs.

II. Hypothalamic Hormones

The peptides released from the hypothalamus to influence the secretion of pituitary polypeptide hormones are known either as the hypophysiotrophic hormones (Buckingham 1978) or as the hypothalamic regulatory hormones (Besser and Mortimer 1974). The latter name emphasizes the fact that they can either stimulate or inhibit the release of pituitary hormones. They are called 'hormones' when their chemical composition is known, and 'factors' when this is as yet unknown even though some of their action is well defined. Thus the first to be demonstrated, which was the factor that causes the pituitary gland to secrete corticotrophin, is still known as the corticotrophin-releasing factor (CRF) because its precise composition is not fully determined. The same applies to the prolactin-releasing factor, and the prolactin release-inhibiting factor. (For the involvement of TRH in the secretion of prolactin, reference may be made to Besser 1974 b and to Buckingham 1978). The other known releasing and inhibiting factors or hormones (Buckingham 1978) are: growth hormone-releasing factor and growth hormone release-inhibiting hormone; melanocyte-stimulating-hormone release-inhibiting factor and melanocyte-releasing factor; TRH; gonadotrophin-releasing hormone, which appears to stimulate the release of both LH and follicle stimulating hormone. (However, TRH influences the release of FSH in men, and of LH in women at the luteal peak: Besser 1974 b.)

It should be noted that various catecholamines, such as dopamine, noradrenaline, indolamine and 5-hydroxytryptamine, have been implicated either as releasing substances or as complementary to known releasing factors or hormones. In a series of studies Buckingham and Hodges (Buckingham 1978) have shown that arginine vasotocin and oxypressin can act as corticotrophin-releasing substances, giving dose-response relationships similar to that shown by hypothalamic extracts. And Buckingham (1979) has demonstrated that corticosterone has a direct influence on the hypothalamus and reduces its capacity both to synthesize and to release CRF.

III. Problems in the Direct Bioassay of Hypophysiotrophic Hormones

As has been discussed in relation to the bioassay of PTH (Chap. 13), one way of developing a new bioassay is first to maintain the target organ in vitro, and observe which enzymatic activities, in which cells, decline as a result of their being deprived of their specific hormone. The relationship between these cells and the specific hormone is then confirmed by adding the hormone to the culture medium and seeing whether, or not, the enzymatic activity is restored. As an exercise, this might have been worth trying for the hypophysiotrophic hormones. Pituitary glands, or segments of these glands, could be maintained in vitro and the effect of deprivation or addition of particular releasing hormones (or factors) could have been studied. However, at the outset of our studies, this did not seem to be a practical approach: first because the histology of the pituitary gland does not apparently lend itself to simple landmarks by which particular cell types can be identified (as, for example, the different tubules of the kidney allow histological mapping of cells influenced by PTH); secondly, because it is by no means clear that a certain cell type in the pituitary gland responds to only one releasing hormone: it is possible that some cell types may produce more than one pituitary hormone and respond to more than one releasing hormone; and thirdly because even if a clear-cut effect could be found in one particular cell type, when one releasing hormone was allowed to act alone on the pituitary gland, it would not have been possible to distinguish the responses of the other cell types when plasma, containing many releasing hormones, was added, as it would be in an assay.

Consequently, a two-step assay system was devised. In this system the releasing hormone, or plasma, is allowed to act on segments of the pituitary gland maintained in vitro, and the amount of pituitary hormone which the releasing hormone causes to be released into the medium is assayed by the appropriate cytochemical bioassay. In this way the high sensitivity of the cytochemical bioassays can be used while magnifying the effect of the releasing hormone through its action on the pituitary gland.

B. Thyrotrophin-Releasing Hormone

I. Structure

Thyrotrophin-releasing hormone, whether human, bovine, ovine or porcine, is a tripeptide containing pyro-glutamate, histidine and prolineamide (e. g. Folkers et al. 1969; Burgus et al. 1970). The ring structure of the pyro-glutamate residue seems to confer some stability on the molecule. In the earlier studies (e. g. Schally et al. 1968, 1969) it was shown to act in vivo in nanogram amounts and to release up to 2000 times its weight of TSH.

II. Distribution of Thyrotrophs in the Pituitary Gland

It first had to be established whether the thyrotrophs were unevenly distributed within the pituitary gland. To test this question, the pituitary gland was cut into four segments, each containing approximately equal amounts of the anterior lobe. Each segment was maintained in vitro for 5 h at 37° C in contact with Trowell's T8 medium, as is done in the other segment assays. At the end of this period, each segment was exposed for 10 min to fresh T8 medium containing TRH (10 ng/ml). The medium was removed and assayed, at $1:10^2$ dilution, for its content of TSH. The results were remarkably constant, the greatest variation being only $\pm 6\%$ of the mean value.

III. Cytochemical Bioassay (Gilbert et al. 1975)

A guinea-pig is killed by asphyxiation with nitrogen; the pituitary gland is removed and quartered to include roughly equal amounts of the anterior lobe in each segment. Each segment is maintained in contact with Trowell's T8 medium for 5 h, as is done in all the segment assays. Two segments are then treated for 10 min with fresh T8 medium containing 0.3% human albumin (free from proteolytic enzymes) and one of two concentrations of TRH; each of the other two segments is exposed to the T8 albumin medium containing one of two concentrations of the plasma which is to be assayed. (The albumin is added to protect any TSH secreted by the segments into the medium.) After 10 min the pituitary tissue is removed and the medium is snap-frozen to −70° C.

It has been found that most, if not all, the secretion of TSH induced by TRH will have occurred by 10 min. The medium removed from each pituitary segment is then assayed, by the cytochemical section assay, for its content of TSH. To measure the TSH released by TRH acting at 100 fg/ml, the medium had to be diluted $1:10^2$

(Gilbert et al. 1975). Thus, depending on the dilution of the plasma, and the subsequent dilution of the medium (the 'effluate' – or 'expectorate', seeing that the segments are not perfused but 'spit-out' TSH), it has readily been possible to measure concentrations of TRH of 10 fg/ml; it is conceivable that lower concentrations can be assayed.

Thyrotrophin-releasing hormone added to the cytochemical bioassay of the secreted TSH had only negligible influence, even at 100 pg/ml. Thus the response was not due to TRH carried over by the 'expectorate'. The T8 albumin medium, lacking plasma or TRH, caused very low secretion of TSH. LH releasing hormone produced only 5% of the amount of TSH stimulated by an equimolar concentration of TRH. Thus by means of this two-stage procedure, it is possible to assay circulating levels of TRH in normal subjects (Fig. 14.1) and in patients with thyroid disease.

C. Corticotrophin-Releasing Factor

I. Cytochemical Bioassay

In this procedure (Buckingham and Hodges 1977 a) the anterior pituitary gland is removed from a decapitated rat (male, albino, Sprague-Dawley) and cut into four approximately equal segments. The segments are incubated together for 2 h at 37° C in 1 ml of an artificial physiological medium (pH 7.4) devised for such purposes by Bradbury et al. (1974); a mixture of 95% oxygen: 5% carbon dioxide is passed constantly through this solution. The segments are then transferred to fresh medium containing lysine-vasopressin (250 μU/ml) and the incubation, under the same gaseous conditions, is continued for a further 15 min. The segments are then separated, each into 1 ml of fresh medium which, for three of the segments, contains the test material; the fourth segment acts as the control. After 15 min exposure to the putative CRF, the segments are removed from the medium, and the latter is assayed by the cytochemical section assay for ACTH.

With extracts of hypothalamus as the source of CRF, the index of precision of ten dose-response regression-lines was 0.046 ± 0.002 (SEM) when the amount of ACTH

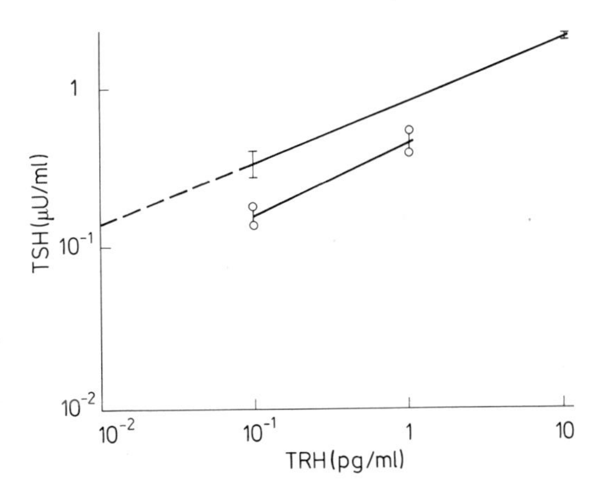

Fig. 14.1. The concentration of TSH secreted into the culture medium by segments of guinea-pig pituitary gland subjected to two concentrations of TRH (*bars*) or to two dilutions of a plasma (*circles*). The bars or circles show the values of each of two dilutions of the effluent for each reading. The responses to the plasma and to the standard preparation are sufficiently parallel for assay

released into the medium was assayed; similar precision was obtained when the ACTH content of the segments was measured.

II. Results with the Bioassay

Many transmitter substances have been tested in vitro for their ability to release ACTH from these adenohypophyseal segments; the most potent were lysine and arginine vasopressin (Buckingham and Hodges 1977 a). Arginine vasotocin gave dose-responses which were parallel to those induced by extracts of hypothalamus (Buckingham and Hodges 1977 a; Buckingham 1978).

Buckingham and Hodges (1977 c) extended this system by isolating hypothalami from rats; incubating them in the artificial medium of Bradbury et al. (1974); and then investigating the effects of various substances on the release of CRF from these hypothalami, and on the synthesis or retention of CRF in the hypothalami after exposure to these substances. The CRF was measured by its effect on segments of the pituitary gland, in vitro, as just described. By these means Buckingham (1979) showed that corticosterone reduced the capacity of the adenohypophysis to release ACTH and of the hypothalamus to release CRF.

It is clear that these techniques open new vistas in the interrelationship between endocrinology and brain physiology and metabolism.

Chapter 15

Prospects: Uses of the Cytochemical Bioassay System

The cytochemical bioassay system involves two types of procedure:

1) The preparation of the target tissue in a form (segment, or section) in which its response to the hormone can be ascertained; this involves maintaining small samples of adult tissue in vitro in a metabolic state which, apart from the effect of stimulation by endogenous hormones, should resemble that which pertains in vivo.
2) The cytochemical measurement of biochemical changes induced by the hormone acting on its target cells. Because cytochemical measurement of such changes is so sensitive, very little material is required, so that only small segments of the target tissue need to be maintained in vitro for such studies.

These two types of procedures, either singly or together, have very wide and very promising applicability. Because the exploitation of these applications lies mainly in the future, they will be discussed only briefly here.

A. Bioassay of Hormones

I. New Assays

There seems every reason for believing that cytochemical bioassays can be developed for any polypeptide hormone for which a sensitive bioassay is needed. It is conceivable that such assays can be also developed for steroid hormones should there be a need for this. The same general principle applies throughout, namely that the hormones must induce some characteristic chemical change in the target cells, and this change should be measurable by modern methods of cytochemistry. The development of such cytochemical bioassays may involve better understanding of how the hormones influence their target cells, or may require the development of new cytochemical techniques. Both can only advance the studies of endocrinology and cytochemistry; neither should be beyond the capabilities of reasonably well-trained research workers.

II. Assay of Similar Factors

There are several examples of molecules, of partially similar amino acid content, which stimulate the same target cells. A simple example is the 1–18 modified, or 1–24 normal sequence of ACTH and the intact 1–39 sequence of the intact hormone molecule (Chap. 3). Another concerns the sialated and desialated glycoprotein hormones (as discussed in Chap. 3).

It may be feasible to discriminate between such similar factors on the basis of the time course of the response which they induce. This has been found to be true for the 1–24 and 1–39 sequences of ACTH (e.g. Chayen et al. 1976). Although equal molar concentrations of both forms of ACTH give similar depletion of ascorbate when measured at the optimal time for the bioassay, the extended time course of the two responses is different (Fig. 15.1). And the time course for the modified 1–18 sequence of ACTH differs slightly from both (Chap. 3).

A somewhat similar phenomenon is the faster response of the thyroid-follicle cells to TSH, as compared with the longer, slower response to thyroid-stimulating immunoglobulins (as discussed in Chap. 10); however, exceptions to this differential rate may occur in samples of plasma having such immunoglobulins of unusually high potency (Loveridge et al. 1979). Why it should be that TSH, and these immunoglobulins, increase the rate of response at high concentrations, remains a mystery; its solution should give useful information concerning the mode of action of hormones.

III. Influence of the 'Biologically Inactive' Part of the Molecule

As shown in Fig. 15.1, the presence of the amino acid sequence 25–39 in the intact molecule of ACTH modified the time course over which the α_{1-39} ACTH molecule acted, in comparison with the 1–24 sequence. One possible interpretation of this time course is that the complete, natural molecule remained attached to the active site longer than did the synthetic 1–24 sequence. This question of 'attachment' becomes of greater consequence when considering the effects of gastrin and related compounds.

The biological activity of the heptadecapeptide gastrin (G_{17}) resides in the first four C-terminal amino acids (as discussed in Chap. 12). Consequently the

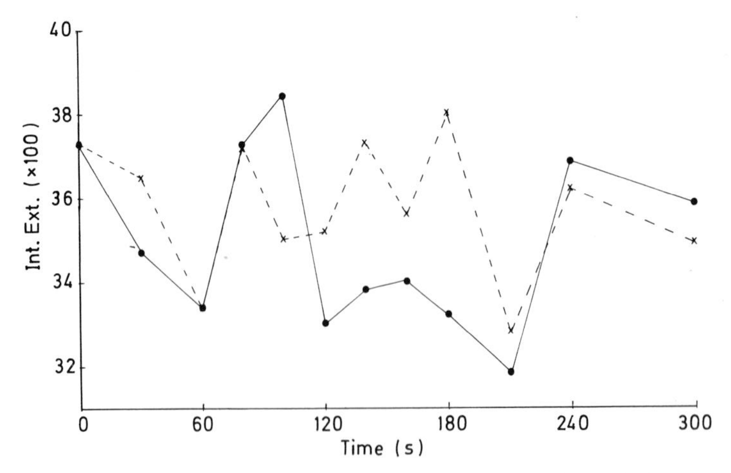

Fig. 15.1. The response time course (Int. Ext. × 100/time) for the synthetic α_{1-24} and the normal α_{1-39} ACTH measured in sections. Both produce the same drop in reducing capacity (ascorbate-depletion) at 60 s, but the response to the α_{1-24} ACTH (*broken line*) has been slower. The secondary response to the α_{1-39} ACTH (*solid line*) is appreciably greater (i.e. between 120 and 200 s) so that the area covered by the whole response (1691 cm^2) is greater than that to the α_{1-24} ACTH (1224 cm^2). (Chayen et al. 1976, p. 67)

commercial pentagastrin (Peptavlon), which contains these critical amino acids, can be used for diagnostic purposes in place of G_{17}. But it is of some concern that these amino acids also occur in CCK-OP and it is possible that this octapeptide might occur in the circulation, as a degradation product of CCK. In as yet unpublished work (by Dr. N. Loveridge, Dr. J. D. Gardner and the present author) it was first shown that, when tested on their own, the natural heptadecapeptide, Peptavlon and the octapeptide (CCK-OP) were equipotent on a molar basis, in the cytochemical section assay for gastrin-like activity (Chap. 12). But when either Peptavlon, or the CCK-OP, was tested in the presence of a low concentration of the natural heptadecapeptide, the response was equal solely to that of the natural heptadecapeptide, even when the other molecule was present at 100 times the molar concentration of the natural hormone. Yet when Peptavlon and the octapeptide were tested together (without G_{17}), the effects were additive.

The immediate applicability of these results would be that G_{17} can be assayed in plasma even if the octapeptide of CCK was also present in the circulation. However, more fundamentally, it seems possible that the biologically inactive region of this hormone may play some part in determining the specificity of the interaction between the heptadecapeptide and its specific target cells. This phenomenon merits investigation in the several other known examples, either where there can be two molecular forms of a hormone, as exemplified by the synthetic α_{1-24} ACTH and the natural α_{1-39} ACTH, or where two different hormones (or hormone-like substances, such as stimulating immunoglobulins) may be available at the same time to act on the same target cells. An example of this type of situation would be in patients with Graves' disease. It has been shown by El Kabir et al. (1971) that the long acting thyroid stimulator can have ACTH-like effects on the adrenal cortex. On the other hand, it is uncommon to find clinical evidence of adrenocortical overactivity in these patients which cannot be ascribed simply to stress. It may be that, in the presence of circulating ACTH, the adrenal gland will not recognize the thyroid-stimulator, just as the parietal cells failed to respond to Peptavlon, or to CCK-OP, in the presence of even a low concentration of the natural G_{17}-gastrin.

IV. Blocking or Stimulating 'Antibodies': The Importance of Parallelism

In Chap. 12 Sect. E.II, it was shown that non-parallel responses caused by samples of plasma merit detailed study: in that instance (in samples from patients with pernicious anaemia), such responses were due to antibodies which appeared to block the ability of the parietal cells to respond to gastrin. It seems likely that other, similar effects will come to light; it is not certain that substances that block the response of the target cells to exogenously added hormone need necessarily be antibodies, or even immunoglobulins. However, when faced with a plasma that produces a non-parallel response, it may be helpful to consider the possibility that there is some blocking molecule, whether it is an antibody to the hormone, or to the receptor (as in the samples from patients with pernicious anaemia), or some other agent which blocks the ability of the cells to respond.

It is also possible that some agents can stimulate the target cells very much as if they were the specific hormone, except that they will produce a dose-response which is not parallel to that induced by the specific hormone. This too merits detailed investigation on those occasions that it is observed.

B. Cytochemistry as Cellular Chemical Pathology

The essence of the sensitivity of the cytochemical bioassays is the fact that cytochemistry is now so far advanced that it can measure changes in the biochemical activity of the target cells when the cells respond to low concentrations of the specific hormone. These changes can be quite small, yet they can be measured and discriminated from changes induced by higher or lower concentrations of the hormone.

This ability to measure metabolic changes, with some degree of sensitivity, can be (and has been) utilized in studying the metabolic changes caused by disease. Because cytochemistry is essentially cellular biochemistry, a small biopsy of the relevant tissue should be sufficient for measuring the biochemical lesion, or the altered metabolism, characteristic of the disease (e. g. Bitensky 1971). At a fairly simple level, Bitensky (1965, 1967) could distinguish between jaundice due to biliary obstruction as against that due to liver cell damage, by examining the localization of alkaline phosphatase in sections from liver biopsies. At a more precise quantitative level, cytochemistry has been able to elucidate cellular changes involved in myocardial dysfunction during prolonged open-heart surgery (Niles et al. 1964; Braimbridge et al. 1973 a, b) and to monitor the state of the myocardium during surgery (e. g. Braimbridge and C̄anković-Darracott 1979). All these evaluations have been done on small drill biopsies taken from the apex of the left ventricle, or on biptome biopsies taken during the course of catheterization.

The detailed analysis of the biochemical alterations induced in rheumatoid arthritis in the cells lining the joints (e. g. Chayen et al. 1971 b; Butcher et al 1973; Henderson et al. 1978, 1979) is an example of how biochemical changes induced by disease can be measured and analysed by cytochemistry. Biopsies of the synovial tissue lining the joints are frequently small, and the relevant cells, the synovial lining cells (or synoviocytes) may constitute only a small fraction of the total tissue. Yet this is one of the advantages of cytochemistry, namely that it relates biochemical activity to histology so that the metabolism of histologically distinct cells can be determined irrespective of the activity of the rest of the tissue sample. Moreover, by maintaining the tissue in vitro, for rather longer than is done for the cytochemical bioassays, and under more rigorously sterile conditions, it is possible to test the therapeutic effect of drugs on the human tissue, so avoiding the dubious extrapolation from what happens in laboratory animals to what might, or might not, occur in the patient.

Another application of cytochemistry has been in the detection of the site at which toxic agents or xenobiotics act, and in measuring the biochemical damage caused by such agents. For example, it is known that liver necrosis caused by certain otherwise innocuous subtances is due to their modification, by the microsomal respiratory system of the liver, into cytotoxic molecules. The fact that these substances cause centrilobular necrosis has been assumed to be due to the activity of this respiratory system being greater in the hepatocytes around the central vein of each lobule. It has now been shown, cytochemically, that the concentration of cytochrome P-450 (the 'marker' of this respiratory system) is greater in the centrilobular than in the periportal hepatocytes (Gooding et al. 1978) and that these cells have a lower concentration of glutathione, which normally protects against these xenobiotics (Smith et al. 1979).

C. Study of Biologically Active Molecules

Just as small pieces of guinea-pig tissue can be maintained in vitro to test their response to hormones in the cytochemical bioassay system, so can the same system be used to investigate (or assay) the effect of other pharmacologically active substances on small biopsies taken from human subjects. Thus, for example, the relative potencies of anti-inflammatory drugs have been determined (Chayen et al. 1970. 1972 b) and studies have been made on the mode of action of glucocorticoids in the synovial tissue of patients with rheumatoid arthritis (Bitensky et al. 1974 b, 1977). In theory, at least, the ethos of the cytochemical bioassay system can be applied to the study of the effect of any biologically active molecules acting on the relevant human tissue, so circumventing the need to rely solely on their influence in laboratory animals. Several other applications of cytochemistry, both in toxicology and in many branches of medical science, have been discussed recently (in Pattison et al. 1979).

D. Conclusion

Cytochemistry has developed rapidly over the past 15 years to become established as a precise form of cellular biochemistry, capable of measuring small and often subtle changes in the metabolic activity of specified cells in a complex tissue. Its major application to date has been in the bioassay of polypeptide hormones. The development of the subject, the bases of the discipline and the cytochemical bioassays currently in use, have been described in this monograph. These cytochemical bioassays are already widely used, especially because of the resurgence of interest in biological (functional) assays. It is likely that new cytochemical bioassays, for other hormones, will be developed in the near future. Since they all utilize the same basic apparatus, and the same expertise, it is not difficult to expand the range of cytochemical bioassays done in any laboratory, once one has become established.

But over and above its use in measuring concentrations of biologically active hormones, the cytochemical bioassay system has great scope in measuring how hormones, and other pharmacologically active agents, produce their effects. And because cytochemistry is now a basically non-disruptive, highly sensitive form of biochemistry at the cellular level, which relates biochemical activity to detailed histology, its scope as a logical extension of conventional biochemistry seems, at present, almost limitless.

Appendix 1

Maintenance Culture

It cannot be over-emphasized that the segments of tissue to be maintained in vitro contain living cells, and that sloppy technique at all stages, especially in washing the utensils, can produce deleterious effects on the viability and responsiveness of the cells. In laboratories newly establishing these procedures, one of the major causes of poor assays has been incorrect washing of the utensils.

A. Cleaning

All the glassware, namely well-slides, Petri dishes, vitreosil dishes, as well as the stainless steel grids and all metal instruments are treated in the following way. They are soaked overnight in an open vessel, kept solely for soaking such material for culture, in tap water, containing Pyroneg (obtained from Diversey Limited, Weston Favell Centre, Northampton NN3 4PD; about six tablespoons in 2 gallons). They are then rinsed well in tap water; then in distilled water and then dried in a cool hot-air oven. They are packaged in small suitable containers, for convenience, and sterilized in a hot-air oven.

The culture outer chambers (Perspex or other plastic material) are cleaned of Lanoline with a dry paper towel, and then are scrubbed with the Pyroneg solution. They are then washed in tap water, rinsed in distilled water, and left to dry in air. They are sterilized by exposure to ultraviolet light just before they are used.

To remove grease, and render the lens tissue compatible with water, pieces of microscope lens tissue are soaked in ethyl ether for 30 min, in a fume cupboard (with no naked flames or sparks). They are then dried in air. This process is done twice again. The dried lens tissue is then washed in three changes of distilled water, each of about 30 min. The lens tissue is allowed to dry in air and then is cut into squares, about 2 cm by 2 cm (i.e. approximately the size of the top of the metal-mesh table). These squares of lens tissue are then sterilized in a hot-air oven.

B. Procedure

One of the aims of this procedure is to get the segments of the target organ into culture as quickly and as gently as possible. The responsiveness of the cells will decrease markedly with clumsy handling, and with time between the death of the animal and being set into the culture pot. The tissue should never be handled with forceps, these being used only on the surrounding or supporting tissue during dissection and cleaning. The tissue should also not be exposed to undue tension or pressure while it is dissected out of the animal. The segments should be in culture within 3–4 min, at the most not more than about 10 min, of the death of the animal.

The animal is killed by asphyxiation in nitrogen or, in the case of rats (e. g. for the LH assay), by cervical dislocation. These procedures avoid the use of agents that may influence membranes. The fur is wetted with water to reduce the loss of hair; the target organ is dissected free and placed in a well-slide which contains a few drops of the culture medium. The well-slide is kept in a Petri dish, and the dish is covered again as soon as the organ is placed in the slide.

In a clean area, preferably under a culture hood, lay the culture pots and vitreosil dishes out with the stainless steel mesh tables placed in them. A square of lens tissue is placed on each mesh-table. With a sterile Pasteur pipette, add the medium to each vitreosil dish so as to soak the lens tissue but not to form a fluid surface above it. (A hole cut in at least one leg of the table avoids air bubbles collecting under the lens tissue.)

With sterile instruments, clean the surrounding supporting tissue, including fat, from the target organ as gently as possible without handling the organ with forceps. Especially for small organs, the scissors and forceps should be small, and kept only for this purpose. It is advisable to check that the scissors are sharp and the forceps efficient; it is best to cut with the scissors held as vertically as possible to avoid a suction effect when they are horizontal to the tissue.

With a sharp, large, sterile blade, attached to a scalpel holder, cut (with the broadest part of the blade) the organ into suitably sized segments. With the flat side of the scalpel blade, transfer each segment to the appropriate lens tissue on its mesh-table. Pick up the vitreosil dish with sterile (large) forceps and transfer it to the culture pot. The lid, which should have had a layer of Lanoline added to its lip prior to killing the animal and before the culture pot is exposed to ultraviolet light for sterilization, is then placed firmly on the culture pot, and turned through a quarter of a circle to spread the Lanoline.

The culture pots, at 35°–37° C, are then attached to the gas line. This must not be made of rubber, which seems selectively to allow diffusion of carbon dioxide. It should be of a firm plastic material. The gas cylinder (5% CO_2 : 95% O_2) is kept at 37° C; the gas is passed through a water jar at 37° C both to ensure that it is at the correct temperature and to humidify it. Gas is allowed to pass, without undue pressure, for 3 min through one port of each culture pot, leaving the other open. The flow of gas out of the open port should be noticeable to the cheek or lip. This second port is then closed, by a clip or spigot.

C. Dilution of the Hormone

With a volumetric bulb pipette, pipette a suitable volume (e. g. 10 ml) of the warmed (37° C) culture medium (e. g. Trowell's T8 medium) into each of a series of stoppered plastic tubes. Withdraw 0.1 ml from the first, and 1 ml from each of the others. (This is expedited by the use of an automatic pipette with disposable plastic tips.) Add 0.1 ml of vehicle containing the appropriate amount of the hormone to the first tube, e. g. to achieve a concentration of 5 pg/ml of ACTH, or 0.1 μU/ml of TSH; replace the stopper and mix well (by inversion at least six times). Because the pH of the Trowell's medium depends on the gaseous phase, it is important to keep the stoppers in the tubes at all times, except when adding or removing fluid. Withdraw 1 ml from the first tube, and add it to the second. Mix well. This gives the first 1 : 10 dilution. Remove 1 ml from the second tube and add it to the third tube, to achieve the next logarithmic dilution; and so on. A new pipette-tip must be used for each dilution. Similar dilutions are made of the samples to be assayed. These dilutions are done close to the end of the 5-h culture, just before the segments are to be exposed to the hormone.

D. Exposure to the Hormone

Remove the lid of the first culture pot. Transfer the vitreosil dish to the bench. With a large pipette, suck out most of the culture medium and discard it. Remove the last few drops from the vitreosil dish with a Pasteur pipette. Start the timing clock instantaneously with pouring the first concentration of the hormone on to the first segment, by pivoting the tube over the forefinger of the other hand. The level of the medium should be just at the level of the mesh-table. When the clock reads either 40 s, or 1 min 40 s, depending on convenience, do the same to the next culture pot, and so on. Exposure to hormone is stopped by lifting the segment off the mesh-table and ejecting it into *n*-hexane (boiling range 67°–70° C, 'free from aromatic hydrocarbons' grade) at −70° C for up to 1 min (not less than 30 s). The chilled segment is then transferred, with cold forceps, to a dry tube at −70° C.

Appendix 2

Microdensitometry with the Vickers M85 Microdensitometer

Ensure that the stage-control stick is in the vertical position. Then place the slide on the microscope table; site the section under a low-power objective and focus the objective. (Note: if the measurements are to be made with the oil-immersion objective, the top lens of the condenser should be lowered slightly and oiled – with no air bubbles – before placing the slide on the table; the condenser should not be in contact with the slide at this stage.) Move the slide by hand so that clear field and the relevant part of the section are reasonably close to the optical axis of the microscope. Secure the section by means of the retaining clips; further movement of the slide will be done by the stage-control stick.

Select the appropriate objective (and add microscope oil if necessary to seal the oil-immersion objective to the slide, again taking care to avoid air bubbles) and obtain the best focus of the specimen. Close the condenser diaphragm; raise or lower the condenser to achieve sharpest focus of the diaphragm in the plane of the section. Centre the condenser. (Note: the image of the diaphragm should be symmetrical as you raise and lower the condenser. If it is not, either light is travelling obliquely, or there is grit in the condenser carriage; remove the condenser, clean it, and replace.) Open the condenser diaphragm fully, checking that the diaphragm is indeed centred by observing the edges of the diaphragm as they cross the boundary of the microscope field.

Close the field diaphragm and render it stable in its mounting by pressing to the left-hand side of its housing. Move the condenser up to give a sharp focus of the field diaphragm in the plane of the section. (With the oil-immersion objective it is necessary to ensure good contact between the oil on the top lens of the condenser and the slide.) This sharp focus should be coincident with the sharpest focus of the specimen. Then open the field diaphragm until it just clears the boundary of the microscope field. (If it is too open, you can have excessive 'glare'.)

Select the size of mask appropriate to the size of cells, or amount of the specimen, to be measured. Locate that mask in the centre of the field. Visualize the immobilized scanning spot (by inserting the reflector in the microscope tube); focus on this spot by means of the eyepieces and not by the microscope focussing screw. Select the correct size of scanning spot, which in virtually every case will be spot 1. Turn the controls to 'automatic', press the integrator button, and ensure that the mask is central in the region scanned by the scanning spot and that the scan fully covers the mask. (This can be achieved by moving the appropriate centring controls of the mask.)

Immobilize the spot and adjust it so that it is in the centre of the mask. Move the section so that the mask lies over clear field (outside the section). Select the required wavelength. Turn the controls to the scanning mode. Set the needle on the transmission density scale to read about 95% transmission. Adjust the gating level to exactly 5 and close the slit width to achieve the smallest possible slit width (for best spectral resolution) compatible with a gating-level of 5, and 95% transmission. Block off the light to the photomultiplier (by the disc at the base of the microscope) to give 'infinite absorption'; adjust the needle, on the transmission density scale, to read infinity. Open the light again, and adjust to 95% transmission, if necessary. Check gating level. This manoeuvre may have to be repeated until the instrument reads infinity and 95% transmission, under the appropriate conditions, with the correct gating level.

Turn controls to automatic and obtain a reading of the integrated relative absorption (density) of the clear field. It will not be zero because you have imposed about 5% absorption on the field, by setting to 95% transmission. The value obtained will have to be subtracted from all values obtained within the section. Then move the slide and measure at least ten appropriate fields.

The linearity of the response of the instrument should have been checked by measuring the integrated relative absorption of a mask against a number of filters of calibrated extinction values. Then the integrated relative absorption (or density) recorded for this size of mask in the specimen can be converted, by means of such a calibration graph, for this particular mask, at this magnification, into absolute units of mean integrated extinction.

Appendix 3

Main Cytochemical Reactions Used in These Assays

A. Reaction for Ascorbate (Chayen et al. 1973b; Alaghband-Zadeh et al. 1973)

This reaction is used in both the ACTH and the LH assays. Solution 1 : 1.35% solution of ferric chloride ($FeCl_3.6H_2O$) in distilled water. Solution 2 : 0.1% solution, in distilled water, of potassium ferricyanide. Immediately before it is to be used, add three volumes of solution 1 to one volume of solution 2, mix well; adjust the pH to 2.4.

Replace the reaction medium with fresh medium after 5–7 min reaction. It may be necessary to react for 14 or 21 min. The sections are then washed in distilled water. They are allowed to dry in air, in the dark, and are then mounted, after a brief rinse in absolute ethanol and in xylene, in DePeX.

B. Naphthylamidase Reaction (Chayen et al. 1973b)

This reaction is used in the assay of thyroid stimulators. The reaction medium must be prepared fresh on the day it is to be used. It contains: 10 ml of an 0.1 *M* acetate buffer, pH 6.1; 8 ml of an 0.85% solution of sodium chloride; 1.0 ml of an 0.02 *M* solution of potassium cyanide (32 mg in 25 ml of water); 1.0 ml of water containing 8 mg of leucine 2-naphthylamide. Just before the medium is to be used, add 10 mg of fast blue B (zinc salt) and adjust the pH to pH 6.5.

The reaction is done in Coplin jars at 37° C. After 15 min, replace the medium with fresh medium. (Again, add the fast blue B salt just before the slides are put into the fresh medium. Hence it is advisable to prepare this medium during the first 15 min reaction.)

It may (rarely) be necessary to repeat this process yet again, if a reaction time of up to 45 min is necessary. Normally 7 min is sufficient for thyroid. After the reaction, rinse the sections in an 0.85% solution of sodium chloride. This can be done at room temperature. Then immerse the sections for 2 min at room temperature in an 0.1 *M* solution of copper sulphate (i.e. 2.5 g $CuSO_4.5H_2O$ in 100 ml of distilled water). Mount the sections in the water-miscible Farrants' medium; it is helpful if the pH of the medium is adjusted to pH 6.5 by the addition of sodium acetate.

C. The Reaction for Carbonic Anhydrase Activity (Loveridge 1978)

This method is used in assaying gastrin-like activity. First, the sections are removed from the cryostat, and left to dry for 10 min, during which time the following solutions are prepared.

Solution 1: 0.1 *M* cobaltous sulphate, 6 ml; 0.5 *M* sulphuric acid, 6 ml; 0.067 *M* potassium dihydrogen orthophosphate, 3 ml; distilled water, 2 ml. Solution 2: 0.75 g sodium hydrogen carbonate (sodium bicarbonate) dissolved in 40 ml of 0.1 *M* HEPES, buffered to pH 7.4 with sodium hydroxide, and containing 0.005% gum tragacanth.

Note: if gum tragacanth is to be used, it should be dissolved on the previous day and left overnight in the buffer. The bicarbonate is added on the day the medium is to be used. Just before the reaction is to be done, add the 17 ml of Solution 1 to the 40 ml of Solution 2. These volumes can be adjusted according to the size of the trough used. The solution is stirred to remove most of the bubbles and used when the pH has reached pH 6.8.

The sections are reacted in this medium, at 37° C, lying flat in a trough which is gently agitated, for 1–3 min. If longer times of reaction are necessary, other aliquots of Solutions 1 and 2 should be mixed during the 3 min reaction and allowed to reach pH 6.8 before the mixture is used to replace the first reaction medium in the trough. Sections should not be reacted in the same mixture of Solutions 1 and 2 for longer than 3 min.

After the reaction, the sections are washed well in running tap water (e. g. for 30s); they are then immersed in a saturated solution of hydrogen sulphide, rinsed in distilled water and mounted in the water-miscible Farrants' medium.

D. The Reaction for Glucose-6-Phosphate Dehydrogenase Activity

(Chayen et al. 1973 b; Chambers et al. 1978 a)

This reaction is used in the assay of PTH. It is done at 37° C. The reaction medium contains: glucose-6-phosphate, 5 m*M*; $NADP^+$, 3 m*M*; phenazine methosulphate, 0.7 m*M*; neotetrazolium chloride (purified),[1] 5 m*M*; potassium cyanide, 10 m*M*. These are dissolved in polyvinyl alcohol in 0.05 *M* glycyl glycine buffer at pH 8.0. Either 12% of the G18/140 Polyviol (polyvinyl alcohol), or preferably 30% of the G04/140 Polyviol (from Wacker-Chemie GMBH, Postfach 1260, 8263 Burghausen/Obb., West Germany) can be used.

The reaction medium is saturated with nitrogen, and excess bubbles are removed. The reaction is done in an atmosphere of nitrogen: the medium is pipetted into a ring that encloses it around the section, and humid nitrogen is passed through the cabinet in which the reaction is done (as described by Chayen et al. 1973 b).

The reaction normally achieves sufficient intensity in 3–5 min. The sections are washed in warm water to assist the removal of the polyvinyl alcohol; they are then allowed to dry in air, and are mounted in Farrants' medium.

The reaction-product is measured at 585 nm.

[1] Most samples of 'neotetrazolium chloride' contain variable amounts of neotetrazolium chloride. It is therefore necessary either to vary the concentration, to allow for impurities in the sample, or to purify the commercial sample by extracting it repeatedly with chloroform, for example in a Soxhlet apparatus.

Appendix 4

Dilution of the Standard Reference Preparations

The reason for including this Appendix is that it may be helpful to know how standard preparations of some hormones have been dissolved and diluted to the levels required for these assays. At such low concentrations all the hormone can be lost by adsorption to the vessels used to contain the solutions, or to pipette aliquots of the solutions. These methods, therefore, are offered as a guide.

It may be noted that it is very advisable to use volumetric glassware for large dilutions. But most hormones should be stored in polystyrene tubes to diminish adsorption, as may occur to glass especially at low concentrations of the hormone.

A. ACTH

The standard generally used in these studies was the powdered form of the α_{1-24} amide (Synacthen), which was most generously donated by CIBA Laboratories, Horsham, West Sussex. I am particularly indebted to the Medical Director, Dr. D. Burley, for his kind readiness to supply me with this preparation.

Dissolve 2 mg in 4 ml of 0.005 *M* hydrochloric acid in a polystyrene tube (concentration: 0.5 mg/ml). Add 0.1 ml of this solution to 9.9 ml of a phosphate-albumin solution (0.01 *M* phosphate buffer, pH 7.4, containing 2.5 mg of protease-free bovine serum albumin per millilitre). This gives a solution containing 5 µg of ACTH per millilitre. Again, take 0.1 ml of this solution and add it to 9.9 ml of the phosphate-albumin solution, to give a concentration of 50 ng/ml.

Aliquots are taken and snap-frozen to −70° C; they are stored in polystyrene tubes at −70° C for up to 3 months. When required, an aliquot is thawed and diluted suitably in Trowell's T8 medium for the assay.

B. Thyrotrophin

The MRC Research A standard, kindly provided by the National Institute for Biological Standards and Control, Hampstead, London, has been used in our studies. Each ampoule contained 50 mU.

The contents of the ampoule were dissolved in 50 ml of a borate buffer (0.13 *M* boric acid; 0.07 *M* sodium hydroxide; to achieve pH 8.0) containing 0.25% protease-free bovine albumin. Thus the concentration of this solution should be 1 mU/ml.

Aliquots of 1 ml were snap-frozen to −70° C and stored at this temperature. The aliquot was dissolved in 99 ml of the borate buffer. (Final concentration of the hormone: 10 µU/ml.) About 100 such subaliquots were prepared and stored at

$-70°$ C in glass tubes that had been rigorously cleaned, rinsed three times in dilute hydrochloric acid and dried, before they were required.

For the section assay, this subaliquot, and the plasma sample, were then diluted, to the required concentrations, in a 'carrier medium' of the following composition: sodium acetate (anhydrous), 410 mg; 1 *M* hydrochloric acid, 1.6 ml; 0.1% gum tragacanth (in distilled water), 20 ml; Trowell's T8 medium deficient in bicarbonate (see Chap. 10. Sect. C.II.), 80 ml. The pH was adjusted to pH 7.6 just before the medium, containing the hormone, was to be used.

C. Gastrin

The reference standard was human synthetic G_{17} I, which has a molecular weight of 2200. It was kindly provided by the National Institute for Biological Standards and Control, Hampstead, London.

The diluent was 0.01 *M* phosphate buffer, pH 7.4, containing protease-free bovine serum albumin at a concentration of 2.5 mg/ml. The contents of a vial (each vial contained 12.6 μg) were dissolved in 1.26 ml of the diluent to achieve a concentration of 10 μg/ml; 0.05 ml of this solution were diluted to 10 ml with diluent, to give a concentration of 50 ng of the hormone per ml. Several aliquots were prepared in polystyrene tubes and snap-frozen to $-70°$ C. They were stored at this temperature for up to 3 months.

D. Parathyroid Hormone

The earlier work done on this hormone used a standard reference preparation of bovine PTH (NIBSC Code 72/286), kindly provided by the National Institute for Biological Standards and Control, Hampstead, London. Each ampoule contained 520 units, or about 200 μg of the hormone.

The contents of the ampoule were dissolved in 0.5 ml of 0.1% acetic acid to give a solution containing approximately 400 μg of the hormone per millilitre. From this solution, 0.1 ml was taken and added to 0.9 ml of hypoparathyroid plasma. (Very out-of-date blood-bank plasma may be used.) This gives a concentration of 40 μg/ml; from such a solution, aliquots of 0.1 ml were taken into polystyrene tubes and snap-frozen to $-70°$ C; they were stored at this temperature.

After thawing, the contents of each tube (0.1 ml) were added to 3.9 ml of Trowell's T8 medium, in order to give a concentration of 1 μg/ml (the new reference standard that we are using, NIBSC Code 77/533, contains only 200 units/ampoule (about 100 μg/ampoule) so that 1.9 ml of Trowell's T8 medium, instead of 3.9 ml, are used). From this solution, 0.1 ml was taken and mixed with 9.9 ml Trowell's T8 medium; 0.1 ml of this solution was added to 9.9 ml of Trowell's medium to achieve a concentration of 100 pg/ml.

From this solution, 0.1 ml was taken and added to 9.8 ml of Trowell's medium plus 0.1 ml of a hypoparathyroid plasma. This gives the top point of the calibration graph, namely 1 pg/ml and 1% plasma, corresponding to a 1/100 dilution of the test plasma. Further dilution, for the full calibration graph, was done in Trowell's medium.

References

Adams DD, Purves HD (1955) A new method of assay for thyrotrophic hormone. Endocrinology 57:17–24

Adams DD, Purves HD (1956) Abnormal responses in the assay of thyrotrophin. Proc Univ Otago Med Sch 34:11–12

Agus ZS, Puschett J.B, Senesky D, Goldberg M (1971) Mode of action of parathyroid hormone and cyclic adenosine 3′, 5′-monophosphate on renal tubular phosphate reabsorption in the dog. J Clin Invest 50:617–626

Agus ZS, Gardner LB, Beck LH, Goldberg M (1973) Effects of parathyroid hormone on renal tubular reabsorption of calcium, sodium and phosphate. Am J Physiol 224:1143–1148

Aikman AA, Wills ED (1974) Studies on lysosomes after irradiation. I. A quantitative histochemical method for the study of lysosomal membrane permeability and acid phosphatase activity. Radiat Res 57:403–415

Alaghband-Zadeh J, Daly JR, Tunbridge RDG, Loveridge N, Chayen J (1972) Methodological refinements in the redox bioassay for adrenocorticotrophin. J. Endocrinol: 58 XIX

Alaghband-Zadeh J, Daly JR, Bitensky L, Chayen J (1974) The cytochemical section assay for corticotrophin. Clin Endocrinol (Oxf) 3:319–327

Allison AC (1968) Lysosomes. In: Bittar EE, Bittar N (eds) The biological basis of medicine vol 3. New York, Academic Press, pp 209–242

Altman FP (1967) PhD dissertation, University of London, p 133

Altman FP (1972) Quantitative dehydrogenase histochemistry with special reference to the pentose shunt dehydrogenases. Prog Histochem Cytochem 4:225–273

Altman FP (1974) Studies on the reduction of tetrazolium salts. III. The products of chemical and enzymic reduction. Histochemistry 38:155–171

Altman FP (1976) Tetrazolium salts and formazans. Prog Histochem Cytochem 9:4–56

Altman FP, Barrnett RJ (1975) The ultra-structural localisation of enzyme activity in unfixed tissue sections. Histochemistry 41:179–183

Altman FP, Butcher RG (1973) Studies on the reduction of tetrazolium salts. I. The isolation and characterisation of a half-formazan intermediate produced during the reduction of neotetrazolium chloride. Histochemie 37:333–350

Altman FP, Chayen J (1965) Retention of nitrogenous material in unfixed sections during incubation for histochemical demonstration of enzymes. Nature 207:1205–1206

Altman FP, Chayen J (1966) The significance of a functioning hydrogen-transport system for the retention of 'soluble' dehydrogenases in unfixed sections. J R Microsc Soc 85:175–180

Altman FP, Bitensky L, Butcher RG, Chayen J Integrated cellular chemistry applied to malignant cells. In: Evans DMD (ed) Cytology automation. Livingstone, Edinburgh, pp 82–99

Altman FP, Moore DS, Chayen J (1975) The direct measurement of cytochrome P–450 in unfixed tissue sections. Histochemistry 41:227–232

Arnaud CD, Goldsmith RS, Bordier PJ, Sizemore GW, Larsen JA, Gilkinson J (1974) Influence of immunoheterogeneity of circulating parathyroid hormone on results of radioimmunoassays of serum in man. Am J Med 56:785–793

Asahina E (1956) The freezing process of plant cell. Contrib Inst Low Temp Sci Hokkaida Univ 10:83–126

Askew AR, Napier BJ, Vinik AI, Terblanche J. (to be published) The bioactivity of different gastrin molecules assessed using a cytochemical section bioassay. Br J Surg

Bacharach U (1975) Cyclic AMP-mediated induction of ornithine decarboxylase of glioma and neuroblastoma cells. Proc Natl Acad Sci USA 72:3087–3091

Bahn RC, Glick D (1954) Studies in histochemistry: Effects of stress conditions, ACTH, cortisone and

desoxycortisone on the quantitative histological distribution of ascorbic acid in adrenal glands of the rat and monkey. Endocrinology 54:672–684
Baker JR (1945) Cytological technique. Methuen, London
Baker JR (1958) Principles of biological microtechnique. Methuen, London
Bangham DR, Butt WR, Hartree AS, Lunenfeld B, Reichert LE, Ross GT, Ryan RJ (1972) Assay of protein hormones related to human reproduction, problem of specificity of assay matter and repressed steroid. Acta Endocrinol (Kbh) 71:625–637
Bangham DR, Berryman I, Burger H, Cotes PM, Furnival BE, Hunter WM, Midgley AR, Mussett MV, Reichert LE, Rosemberg E, Ryan RJ, Wide L (1973) An international collaborative study of 69/104, a reference preparation of human pituitary FSH and LH. J Clin Endocrinol Metab 36:647–660
Barka T, Anderson PJ (1963) Histochemistry. Harper & Row, New York
Barter R, Danielli JF, Davies HG (1955) A quantitative cytochemical method for demonstrating alkaline phosphatase activity. Proc R Soc Lond (Biol) 144:412
Belchetz PE, Elkeles RS (1976) Idiopathic hypopituitarism with biologically inactive TSH. Proc R Soc Med 69:428–429
Bell LGE (1956) Freeze-drying. In: Oster G, Pollister AW (eds) Physical techniques in biological research, vol 3. Academic Press, New York, pp 1–27
Berson SA, Yalow RS (1971) Nature of immunoreactive gastrin extracted from tissues of gastrointestinal tract. Gastroenterology 60:215–222
Besser GM (1974 a) Hypothalamus as an endocrine organ. I. Br Med J 3:560–564
Besser G (1974 b) Hypothalamus as an endocrine organ. II. Br Med J 3:613–615
Besser GM (1977) Pituitary and hypothalamic physiology. In: McGowan GK, Walters G (eds) Hypothalamic and pituitary hormones. J Clin Pathol [Suppl 7] 30:8–11
Besser GM, Mortimer CH (1974) Hypothalamic regulatory hormones: a review. J Clin Pathol 27:173–184
Besser GM, Cullen DR, Irvine WJ, Ratcliffe JG, Landon J (1971) Immunoreactive corticotrophin levels in adrenocortical insufficiency. Br med J 1:374
Bitensky L (1963 a) The reversible activation of lysosomes in normal cells and the effects of pathological conditions. In: De Reuck AVS, Cameron MP (eds) Ciba Foundation symposium on lysosomes. Churchill, London, pp 362–375
Bitensky L (1963 b) Cytotoxic action of antibodies. Br Med Bull 19:241–244
Bitensky L (1965) Enzyme histochemistry as an aid to diagnosis. In: Ruyssen, HA Vandendriessche B (eds) Enzymes in clinical chemistry. Elsevier, Amsterdam, pp 62–66
Bitensky L (1967) Histochemistry of liver disease. In: Read AE (ed) The liver. Butterworths, London, pp 61–72
Bitensky L (1971) Histochemistry as a diagnostic tool. Int Pathol Bull 2:7–15
Bitensky L, Chayen J (1977) Histochemical methods for the study of lysosomes. In: Dingle JT (ed) Lysosomes, a laboratory handbook 2nd edn. North Holland, Amsterdam, pp 209–243
Bitensky L, Butcher RG, Chayen J (1973) Quantitative cytochemistry in the study of lysosomal function. In: Dingle JT (ed) Lysosomes in biology and pathology vol 3. North Holland, Amsterdam, pp 465–510
Bitensky L, Alaghband-Zadeh J, Chayen J (1974 a) Studies on thyroid stimulating hormone and the long-acting thyroid stimulating hormone. Clin Endocrinol (Oxf) 3:363–374
Bitensky L, Butcher RG, Johnstone J, Chayen J (1974 b) Effect of glucocorticoids on lysosomes in synovial lining cells in rheumatoid arthritis. Ann Rheum Dis 33:57–61
Bitensky L, Gilbert DM, Chayen J (1976) The cytochemical investigation of hormonal stimulation of the thyroid and its application in highly sensitive bioassay systems. In: Von zur Mühlen A, Schleusener H (eds) Biochemical basis of thyroid stimulation and thyroid hormone action. Thieme, Stuttgart, pp 115–127
Bitensky L, Cashman B, Johnstone J, Chayen J (1977) Effect of glucocorticoids on the hexose monophosphate pathway in human rheumatoid synovial lining cells in vitro and in vivo. Ann Rheum Dis 36:448–452
Bitensky L, Besser GM, Ealey PA, Gilbert DM, Chayen R, Chayen J (1978 a) Possible involvement of polyamines in the response of thyroid follicle cells to thyrotrophin. J Endocrinol 77:29–30 P
Bitensky L, Ealey PA, Chayen R, Chayen J (1978 b) Spermidine-mediated response to thyrotropin. Proceedings of the Endocrine Society (USA), June 1978, Miami, Abs. 325
Bitensky L, Loveridge N, Chayen J, Gardner JD, Bottazzo GF, Doniach D (1979) Inhibition of gastrin-responsiveness by parietal cell antibodies. Clin Sci Mol Med 56:17 P

Bittar EE (1964) Cell pH. Butterworths, London
Blair EL, Wood DD (1968) The estimation of gastrin activity in blood. J. Physiol (Lond) 194:44–45 P
Bloomfield GA, Holdaway IM, Corrin B, Ratcliffe JG, Rees GM, Ellison M, Rees LH (1977) Lung tumours and ACTH production. Clin Endocrinol (Oxf) 6:95–104
Bogdanove EM, Campbell GT, Blair ED, Mola ME, Miller AE, Grossman GH (1974) Gonad-pituitary feedback involves qualitation change: Androgens alter the type of FSH secreted by the rat pituitary. Endocrinology 95:219–228
Boivin A, Vendrely R, Vendrely C (1948) L'acide désoxyribonucleique du noyau cellulaire, dépositaire des caractères héreditaires: arguments d'ordre analytique. CR Acad Sci [D] (Paris) 226:1061–1063
Borth R (1952) The chromatographic method for the determination of urinary pregnanediol. Ciba Found Colloq Endocrinol 2:45–57
Brachet J (1950) Chemical embryology. Interscience, New York London
Brachet J (1954) The use of basic dyes and ribonuclease for the cytochemical detection of ribonucleic acid. Q J Microsc Sci 94:1–10
Bradbury MWB, Burden J, Hillhouse EW, Jones MT (1974) Stimulation electrically and by acetylcholine of the rat hypothalamus in vitro. J Physiol (Lond) 239:269–283
Braimbridge MV, Ćanković-Darracott S (1979) Quantitative polarization microscopy and cytochemistry in assessing myocardial function. In: Pattison JR, Bitensky L, Chayen J (eds) Quantitative cytochemistry and its applications. Academic Press, London 221–230
Braimbridge MV, Darracott S, Bitensky L, Chayen J (1973 a) Cytochemical analysis of left ventricular biopsies in open-heart surgery: a pilot study. Beitr Pathol 148:255–264
Braimbridge MV, Darracott S, Clement AJ, Bitensky L, Chayen J (1973 b)Myocardial deterioration during aortic valve replacement assessed by cellular biological tests. J Thorac Cardiovasc Surg 66:241–246
British Pharmacopoeia (1963), p. 1087. The Pharmaceutical Press, London
Brown JR (1959) The measurement of thyroid stimulating hormone (TSH) in body fluids: a critical review. Acta Endocrinol (Kbh) 32:289–309
Brown JR, Munro DS (1967) A new in vitro assay for thyroid stimulating hormone. J Endocrinol 38:439–445
Brown PS, Wells M, Cunningham FJ (1964) A method for studying the mode of action of oral contraceptives. Lancet 2:446–447
Buckingham JC (1974) PhD dissertation University of London
Buckingham JC (1977) The endocrine function of the hypothalamus. J Pharm Pharmacol 29:649–656
Buckingham JC (1978) The hypophysiotrophic hormones. Prog Med Chem 15:165–210
Buckingham JC (1979) The influence of corticosteroids on the secretion of corticotrophin and its hypothalamic releasing hormone. J Physiol (Lond) 286:331–342
Buckingham JC, Hodges JR (1974) Interrelationships of pituitary and plasma corticotrophin and plasma corticosterone in adrenalectomized and stressed, adrenalectomized rats. J. Endocrinol 63:213–222
Buckingham JC, Hodges JR (1975) Interrelationships of pituitary and plasma corticotrophin and plasma corticosterone during adrenocortical regeneration in the rat. J Endocrinol 67:411–417
Buckingham JC, Hodges JR (1976) Hypothalamo-pituitary adrenocortical function in the rat after treatment with betamethasone. Br J Pharmacol 56:235–239
Buckingham JC, Hodges JR (1977 a) The use of corticotrophin production by adenohypophyseal tissue in vitro for the detection and estimation of potential corticotrophin releasing factors. J Endocrinol 72:187–193
Buckingham JC, Hodges JR (1977 b) Functional activity of the hypothalamo-pituitary complex in the rat after betamethasone treatment. J Endocrinol 74:297–302
Buckingham JC, Hodges JR (1977c) Production of corticotrophin releasing hormone by the isolated hypothalamus of the rat. J Physiol (Lond) 272:469–479
Buckingham JC, Chayen J, Hodges JR, Robertson WR, Weisz J (to be published) A cytochemical section assay method for the determination of luteinizing hormone. J Endocrinol
Bungenberg de Jong HG (1936) La coacervation. Les coacervates et leur importance en biologie. Actual Sci Ind 1:397–412
Burgus R, Dunn TF, Desiderio D, Ward DN, Vale W, Guillemin R (1970) Characterization of ovine hypothalamic hypophysiotrophic TSH-releasing factor. Nature 226:321–325
Butcher RG (1970) Studies on succinate oxidation. The use of intact sections. Exp Cell Res 60:54–60
Butcher RG (1971 a) The chemical determination of section thickness. Histochemie 28:131–136

Butcher RG (1971 b) Tissue stabilisation during histochemical reactions: the use of collagen polypeptides. Histochemie 28:231–235

Butcher RG (1972) Precise cytochemical measurement of neotetrazolium formazan by scanning and integrating microdensitometry. Histochemie 32:171–190

Butcher RG, Altman FP (1973) Studies on the reduction of tetrazolium salts. II. The measurement of the half reduced and fully reduced formazans of neotetrazolium chloride in tissue sections. Histochemie 37:351–363

Butcher RG, Chayen J (1966) Quantitative studies on the alkaline phosphatase reaction. J R Microsc Soc 85:111–117

Butcher RG, Bitensky L, Cashman B, Chayen J (1973) Differences in the redox balance in human rheumatoid and non-rheumatoid synovial lining cells. Beitr Pathol 148:265–274

Byrnes DJ, Young JD, Chisholm DJ, Lazarus L (1970) Serum gastrin in patients with peptic ulceration. Br Med J 2:626–629

Byus CV, Russell DH (1975 a) Ornithine decarboxylase activity: control by cyclic nucleotides. Science 187:650–652

Byus CV, Russell DH (1975 b) Effects of methyl xanthine derivatives on cyclic AMP levels and ornithine decarboxylase activity of rat tissues. Life Sci 15:1991–1997

Caspersson T (1940) Methods for determination of absorption spectra of cell structures. J R Microsc Soc 60:8–25

Caspersson T (1947) Relations between nucleic acid and protein synthesis. Symp Soc Exp Biol 1:127–151

Caspersson T (1950) Cell growth and cell function. Norton, New York

Challand G, Goldie D, Landon J (1974) Immunoassay in the diagnostic laboratory. Br Med Bull 30:38–43

Chambers DJ (1979) PhD dissertation Brunel University

Chambers DJ, Chayen J (1976) The response of a plasma membrane enzyme to very low concentrations of corticotrophin in the cytochemical section bioassay system. J Endocrinol 68:24 P

Chambers DJ, Zanelli JM, Parsons JA, Chayen J (1976) Cytochemical responses of guinea-pig kidney cortex to low concentrations of bovine parathyroid hormone (0.001–0.1 pg/ml). J Endocrinol 71:87 P

Chambers DJ, Dunham J, Zanelli JM, Parsons JA, Bitensky L, Chayen J (1978 a) A sensitive bioassay of parathyroid hormone in plasma. Clin Endocrinol (Oxf) 9:375–379

Chambers DJ, Schäfer H, Laugharn JA Jr, Johnstone J, Zanelli JM, Parsons JA, Bitensky L, Chayen J (1978 b) Dose-related activation by PTH of specific enzymes in different regions of the kidney. In: Copp DH, Talmage RV (eds) Endocrinology of calcium metabolism. Excerpta Medica, Amsterdam Oxford, pp 216–220

Chayen J (1953) Ascorbic acid and its intracellular localization, with special reference to plants. Int Rev Cytol 2:78–131

Chayen J (1967) Interference microscopy. In: Bourne GH (ed) In vivo techniques in histology. Williams & Wilkins, Baltimore, pp 40–68

Chayen J (1978 a) The cytochemical approach to hormone assay. Int Rev Cytol 53:333–396

Chayen J (1978 b) Microdensitometry. In: Slater TF (ed) Biochemical mechanisms of liver injury. Academic Press, London New York, pp 257–291

Chayen J, Bitensky L (1968) Multiphase biochemistry. In: Bittar EE, Bittar N (eds) The biological basis of medicine, vol 1. Academic Press, New York, pp 337–368

Chayen J, Denby E (1960) The distribution of deoxyribonucleic acid in homogenates of plant roots. Exp Cell Res 20:182–197

Chayen J, Denby EF (1968) Biophysical technique as applied to cell biology. Methuen, London

Chayen J, Jackson SF (1957) Cytoplasmic particles of plant roots. Symp Soc Exp Biol 10:134–149

Chayen J, Miles UJ (1953) The preservation and investigation of plant mitochondria. Q J Microsc Sci 94:29–35

Chayen J, Norris KP (1953) Cytoplasmic localization of nucleic acids in plant cells. Nature 171:472

Chayen R, Roberts ER (1955) Some observations on the metachromatic reaction. Sci J R Coll Sci 25:50–56

Chayen J, Bitensky L, Wells PJ (1966) Mitochondrial enzyme latency and its significance in histochemistry and biochemistry. J R Microsc Soc 86:69–74

Chayen J, Bitensky L, Butcher RG, Poulter LW, Ubhi GS (1970) Methods for the direct measurement of anti-inflamatory action on human tissue maintained in vitro. Br J Dermatol [Suppl. 6] 82: 62–81

Chayen J, Loveridge N, Daly JR (1971a) The measurable effect of low concentrations (pg/ml) of ACTH acting on reducing groups of adrenal cortex maintained in organ culture. Clin Sci 41:2 P
Chayen J, Bitensky L, Butcher RG, Cashman B (1971 b) Evidence for altered lysosomal membranes in synovial lining cells from human rheumatoid joints. Beitr Pathol 142:137–149
Chayen J, Loveridge N, Daly JR (1972a) A sensitive bioassay for adrenocorticotrophic hormone in human plasma. Clin Endocrinol (Oxf) 1:219–233
Chayen J, Bitensky L, Ubhi GS (1972 b) The experimental modification of lysosomal dysfunction by anti-inflammatory drugs acting in vitro. Beitr Pathol 147:6–20
Chayen J, Altman FP, Butcher RG (1973 a) The effect of certain drugs on the production and possible utilization of reducing equivalents outside the mitochondria. In: Dikstein S (ed) Fundamentals of cell pharmacology. Thomas, Springfield, pp 196–230
Chayen J, Bitensky L, Butcher RG (1973 b) Practical histochemistry. Wiley, London New York
Chayen J, Loveridge N, Ubhi GS (1973 c) The use of menadione as an intermediate hydrogen carrier for measuring cytoplasmic dehydrogenating enzyme activities. Histochemie 35:75–80
Chayen J, Bitensky L, Butcher RG, Altman FP (1974a) Cellular biochemical assessment of steroid activity. Adv Steroid Biochem Pharmacol 4:1–60
Chayen J, Bitensky L, Chambers DJ, Loveridge N, Daly JR (1974b) Studies on the mechanism of cytochemical bioassays. Clin Endocrinol (Oxf) 3:349–360
Chayen J, Bitensky L, Butcher RG (1975) Histochemie: Grundlagen und Methoden. Verlag Chemie, Weinheim
Chayen J, Daly JR, Loveridge N, Bitensky L (1976) The cytochemical bioassay of hormones. Recent Prog Horm Res 32:33–79
Chayen J, Chayen R, Besser GM, Bitensky L (1979) Quantitative cytochemistry in the detection and analysis of sensitive, rapid effects of hormones. Biochem Soc Trans: 857–860
Chèvremont M, Frederic J (1943) Une nouvelle méthode histochimique de mise en évidence des substances à fraction sulfhydrile. Arch Biol (Liege) 54:589–605
Chèvremont M, Firket H (1953) Alkaline phosphatase of the nucleus. Int Rev Cytol 2:261–288
Christiansen E, Magid E (1970) Effects of phosphate, HEPES, N_2O and CO on the kinetics of human erythrocyte carbonic anhydrase B and C. Biochim Biophys Acta 220:630–632
Christofides ND, Modlin I, Sarson DL, Albuquerque RH, Ghatei MA, Adrian TE, Bloom SR (1978) The release of gastrin, PP, VIP and motilin after an oral water load and atropine in man. Clin Sci Mol Med 54:12–13 P
Coleman JE (1967) Mechanism of action of carbonic anhydrase: substrate, sulfonamide, and anion binding. J Biol Chem 242:5212–5219
Cornfield J (1970) Invalidity in bioassays. In: McArthur JW, Colton T (eds) Statistics in endocrinology. MIT Cambridge, pp 145–162
Corti A, Dave C, Williams-Ashman HG, Mihich E, Schenone A (1974) Specific inhibition of enzymic decarboxylation of S-adenosylmethionine by methylglyoxal bis (guanyl hydrazone) and related substances. Biochem J 139:351–357
Coulton L (1977) Temporal relationship between glucose 6-phosphate dehydrogenase activity and DNA-synthesis. Histochemistry 50:207–215
Daly JR (1977) Assays for thyroid stimulating hormone. In: Herrmann J, Krüskemper HL, Weinheimer B (eds) Schilddrüse 1975. Thieme, Stuttgart, pp 208–216
Daly JR, Evans JI (1974) Daily rhythms of steroid and associated pituitary hormones in man and their relation to sleep. Adv Steroid Biochem Pharmacol 4:61–110
Daly JR, Bitensky L, Loveridge N, Chayen J (1972) Preliminary assessment of a highly sensitive bioassay 'in vitro' for ACTH. J Endocrinol 53: XXXIV
Daly JR, Fleisher MR, Chambers DJ, Bitensky L, Chayen J (1974 a) Application of the cytochemical bioassay for corticotrophin to clinical and physiological studies in man. Clin Endocrinol (Oxf) 3:335–345
Daly JR, Fletcher MR, Glass D, Chambers DJ, Bitensky L, Chayen J (1974 b) Comparison of effects of long-term corticotrophin and corticosteroid treatment on responses of plasma growth hormone, ACTH and corticosteroid to hypoglycaemia. Br Med J ii:521–524
Daly JR, Loveridge N, Bitensky L, Chayen J (1974 c) The cytochemical bioassay of corticotrophin. Clin Endocrinol (Oxf) 3:311–318
Daly JR, Alaghband-Zadeh J, Loveridge N, Chayen J (1977) The cytochemical bioassay of corticotrophin (ACTH). Ann NY Acad Sci 297:242–258
Danielli JF (1953) Cytochemistry: a critical approach. Wiley, New York

Danielli JF (1958) The calcium phosphate precipitation method for alkaline phosphatase. In: Danielli JF (ed) General cytochemical methods, vol 1, Academic Press, New York, pp 423-443

Davenport HW (1962) Carbonic anhydrase inhibition and physiological function. In: Mongar JL, de Reuck AVS (eds) Enzymes and drug action. Churchill, London, pp 16–29

Davies HG (1954) The action of fixatives on the ultra-violet-absorbing components of chick fibroblasts. Q J Microsc Sci 95:443–457

Davies HG (1958) The determination of mass and concentration by microscope interferometry. In: Danielli JF (ed) General cytochemical methods, vol 1. Academic Press, New York, pp 55–161

Davies HG, Wilkins MHF, Chayen J, La Cour LF (1954) The use of the interference microscope to determine dry mass in living cells and as a quantitative cytochemical method. Q J Microsc Sci 95:271–304

de Duve C (1959) Lysosomes, a new group of cytoplasmic particles. In: Hayashi T (ed) Subcellular particles. Ronald New York, pp 128–159

de Duve C (1963) The lysosome concept. In: De Reuck AVS, Cameron MP (eds) Ciba foundation symposium on lysosomes. Churchill, London, pp 1–35

de Duve C (1969) The lysosome in retrospect. In: Dingle JT, Fell HB (eds) Lysosomes in biology and pathology, vol 1. North Holland, Amsterdam pp 3–40

Deeley EM (1955) An integrating microdensitometer for biological cells. J Sci Instrum 32:263–267

Deeley EM, Richards BM, Walker PMB, Davies HG (1954) Measurements of Feulgen stain during the cell-cycle with a new photo-electric scanning device. Exp Cell Res 6:569–572

Deeley EM, Davies HG, Chayen J (1957) The DNA content of cells in the root of *Vica faba*. Exp Cell Res 12:582–591

Desplan C, Julienne A, Moukhtar MS, Raulais D, Rivaille P, Milhaud G (1978) The isolation of circulating immunoreactive human parathyroid hormone: comparison with a 1–34 N-terminal synthetic fragment. Proc Soc Exp Biol Med 157:241–244

Diczfalusy E (1954) An improved method for the bioassay of chorionic gonadotrophin. Acta Endocrinol (Kbh) 17:58–73

Diebel ND, Yamamoto M, Bogdanove EM (1973) Discrepancies between radioimmunoassays and bioassay for rat FSH: Evidence that androgen treatment and withdrawal can alter bioassay-immunoassay ratios. Endocrinology 92:1065–1078

Diem K, Lentner C (eds) (1970) Documenta Geigy, 7th edn. Ciba-Geigy, Basle, p 575

Dingle JT, Barrett AJ (1969) Some special methods for the investigation of the lysosomal system. In: Dingle JT, Fell HB (eds) Lysosomes in biology and pathology, vol 2. North Holland, Amsterdam, pp 555–566

Dixon M, Webb EC (1964) Enzymes, 2nd edn. Longmans Green, London

Dockray GJ, Taylor IL (1976) Heptadecapeptide gastrin: Measurement in blood by specific radioimmunoassay. Gastroenterology 71:971–977

Dockray GJ, Walsh JH (1975) Amino terminal gastrin fragment in serum of Zollinger-Ellison syndrome patients. Gastroenterology 68:222–230

Döhler KD, Hashimoto T, von zur Mühlen A (1977) Cytochemical assay of thyroid stimulating hormone (TSH). GIT Fachz Lab 21:300–303

Döhler KD, Hashimoto T, von zur Mühlen A (1978) Use of a cytochemical bioassay for determination of thyroid stimulating hormone (TSH) in clinical investigation. International symposium on radioimmunoassay and related procedures in medicine. International Atomic Energy Agency, World Health Organization pp 1–18 Vienna

Donald RA (1971) Plasma immunoreactive corticotrophin and cortisol response to insulin hypoglycaemia in normal subjects and patients with pituitary disease. J Clin Endocrinol Metab 32:225–231

Dufau M, Catt K (1975) A highly sensitive bioassay for LH and hCG in human plasma. Acta Endocrinol [Suppl] (Kbh) 199:189

Dufau ML, Catt KJ, Tsuruhara T (1971) Gonadotropin stimulation of testosterone production by the rat testis in vitro. Biochim Biophys Acta 252:574–579

Dufau M, Mendelson CR, Catt KJ (1974) A highly sensitive in vitro bioassay for luteinizing hormone and chorionic gonadotropin: testosterone production by dispersed Leydig cells. J Clin Endocrinol Metab 39:610–617

Dufau M, Beitins IZ, McArthur JW, Catt K (1976 a) Effects of luteinizing hormone releasing hormone (LHRH) upon bioactive and immunoreactive serum LH levels in normal subjects. J Clin Endocrinol Metab 43:658–667

Dufau M, Pock R, Neubauer A, Catt K (1976b) In vitro bioassay of LH in human serum: The rat interstitial cell testosterone (RICT) assay. J Clin Endocrinol Metab 42:958–968
Dumonde DC, Walter CM, Bitensky L, Cunningham GJ, Chayen J (1961) Intracellular response to an iso-immune reaction at the surface of ascites tumour cells. Nature 192:1302
Dumonde DC, Bitensky L, Cunningham GJ, Chayen J (1965) The effects of antibodies on cells. I. Biochemical and histochemical effects of antibodies and complement on ascites tumour cells. Immunology 8:25–36
Dumont JE (1971) The action of thyrotropin on thyroid metabolism. Vitam Horm 29:287–412
Dunham J, Shedden RG, Catterall A, Bitensky L, Chayen J (1977) Pentose-shunt oxidation in the periosteal cells in healing fractures. Calcif Tissue Res 23:77–81
Ealey PA (1978) Further validation of the cytochemical segment assay of thyrotrophin. J Endocrinol 79:57 P
Ealey PA (1979) PhD dissertation, University of London
Eggleston LV, Krebs HA (1974) Regulation of the pentose phosphate cycle. Biochem J 138:425–435
Ekins RP (1974) Basic principles and theory. Br Med Bull 30:3–11
Ekins RP (1976) General principles of hormone assay. In: Loraine JA, Bell ET (eds) Hormone assays and their clinical application, 2nd edn. Livingstone, London Edinburgh
El Kabir DJ, Hockaday TDR, Richards MR, Dandona P, Naftolin F (1971) Some effects of sera containing long-acting thyroid stimulator on the adrenal cortex of the mouse. Proc R Soc Med 64:154–155
Engström A (1946) Quantitative micro- and histochemical elementary analysis by Roentgen absorption spectrography. Acta Radiol [Suppl] (Stockh) 63:1–106
Engström A (1962) X-ray microanalysis in biology and medicine. Elsevier, Amsterdam
Engström A, Lindström B (1958) The weighing of cellular structures by ultrasoft X-rays. In: Danielli JF (ed) General cytochemical methods, vol 1. Academic Press, New York pp 1–53
Erhardt FW, Hashimoto T (1977) Biological and immunological activities of high-molecular weight TSH from human pituitaries. Acta Endocrinol [Suppl 208] (Kbh) 84:1–2
Erhardt FW, Scriba PC (1977) High molecular thyrotrophin ('big'-TSH) from human pituitaries: preparation and partial characterization. Acta Endocrinol (Kbh) 85:698–712
European Pharmacopoeia (1971) Statistical analysis of results of biological assays and tests. In: European Pharmacopoeia, vol II, Maisonneuve, Paris, pp 441–498
European PTH Study Group (1978) Interlaboratory comparison of radioimmunological parathyroid hormone determination. Eur J Clin Invest 8:149–154
Faglia G, Ferrari C, Paracchi A, Spada A, Beck-Peccoz P (1975) Triiodothyronine release to thyrotrophin releasing hormone in patients with hypothalamic-pituitary disorders. Clin Endocrinol (Oxf) 4:585–590
Faglia G, Bitensky L, Pinchera A, Ferrari C, Paracchi A, Beck-Peccoz P, Spada A (1979) Thyrotropin secretion in patients with central hypothyroidism: evidence for reduced biological activity of immunoreactive TSH. J Clin Endocrinol Metab, 48:989–998
Fell HB (1969) Heberden Oration, 1968: Role of biological membranes in some skeletal reactions. Ann Rheum Dis 28:213–227
Fell HB, Weiss L (1965) The effect of antiserum alone and with hydrocortisone on foetal mouse bones in culture. J Exp Med 121:551–556
Fenton S, Somers S, Heath DA (1978) Preliminary studies with the sensitive cytochemical assay for parathyroid hormone. Clin Endocrinol (Oxf) 9:381–384
Feulgen R, Rossenbeck H (1924) Mikroskopisch-chemischer Nachweis einer Nucleinsäure vom Typus der Thymonucleinsäure und die darauf beruhende elective Färbung von Zellkernen in mikroskopischen Präparaten. Z Physiol Chem 135:203–248
Filipusson H, Hornby WE (1970) The preparation and properties of yeast β-fructofuranosidase chemically attached to polystyrene. Biochem J 120:215–219
Fleisher MR, Glass D, Bitensky L, Chayen J, Daly JR (1974) Plasma corticotrophin levels during insulin hypoglycaemia: comparison of radioimmunoassay and cytochemical bioassay. Clin Endocrinol (Oxf) 3:203–208
Folkers K, Enzmann F, Bøler J, Bowers CY, Schally AV (1969) Discovery of modification of the synthetic tripeptide-sequence of the thyrotropin releasing hormone having activity. Biochem Biophys Res Commun 37:123–126
Fraser RD, Chayen J (1952) The detection of nucleic acids in tissues by infra-red microspectometry. Exp Cell Res 3:492–493

Fujita T, Uezu A, Ota K, Ohata M (1977) Control of parathyroid hormone hydrolysis by the kidney. Contrib Nephrol 6:136–140

Fukushima M, Stevens VC, Gantt CL, Vorys N (1964) Urinary FSH and LH excretion during the normal menstrual cycle. J Clin Endocrinol Metab 24:205–213

Gaddum JH (1953) Bioassays and mathematics. Pharmacol Rev. 5:87–134

Gahan PB (1962) The selection of a suitable control for studies on liver regeneration. Biochem J 84:118 P

Gaillard PJ (1955) Parathyroid gland tissue and bone in vitro. Exp Cell Res [Suppl] 3, 154–169

Gaillard PJ (1961) Parathyroid and bone tissue in culture. In: Greep RO, Talmage RV (eds) The parathyroids. Thomas, Springfield, pp 20–48

Gaillard PJ, Schaberg A (1965) Endocrine glands. In: Willmer EN (ed) Cells and tissues in culture, vol. 2. Academic Press, London New York, pp 631–695

Gallagher TF, Hellman L, Finkelstein J, Yoshida K, Weitzman ED, Roffwarg HD, Fukushima DK (1972) Hyperthyroidism and cortisol secretion in man. J Clin Endocrinol Metab 34:919–927

Gallagher TF, Yoshida K, Roffwarg HD, Fukushima DK, Weitzman ED, Hellman L (1973) ACTH and cortisol secretory patterns in man. J Clin Endocrinol Metab 36:1058–1068

Garren LD, Gill GN, Masui H, Walton GM (1971) On the mechanism of action of ACTH. Recent Prog Horm Res 27:433–475

Gewirtz G, Yalow RS (1974) Ectopic ACTH production in carcinoma of the lung. J Clin Invest 53:1022–1032

Gewirtz G, Schneider B, Krieger DT, Yalow RS (1974) Big ACTH: Conversion to biologically active ACTH by trypsin. J Clin Endocrinol Metab 38:227–230

Gilbert DM, Bitensky L, Besser GM, Chayen J (1975) Measurable effects of low concentrations (100 fg/ml) of thyrotrophin releasing hormone. J Endocrinol 64: 25 P

Gilbert DM, Besser GM, Bitensky L, Chayen J (1977 a) Development of a cytochemical section-assay for thyroid stimulators. J Endocrinol 75:40 P

Gilbert DM, Besser GM, Bitensky L, Chayen J (1977 b) The effect of spermidine on lysosomes in thyroid follicle cells. Biochem Soc Trans 5:1063–1064

Gillette JR, Conney AH, Cosmides GJ, Estabrook RW, Fouts JR, Mannering GJ (eds) (1969) Microsomes and drug oxidations. Academic Press, New York London

Glick D (1962) Quantitative chemical techniques of histo- and cytochemistry, vol 1. Wiley, New York

Glick D (1963) Quantitative chemical techniques of histo- and cytochemistry, vol 2. Wiley, New York

Glick D (1967) Usage of 'histochemical', 'staining' and 'biochemical' in histochemical literature. J Histochem Cytochem 15:299

Glick D, Engström A, Malmström BG (1951) A critical evaluation of quantitative histo- and cytochemical microscopic techniques. Science 114:253–258

Goldstein DJ (1970) Aspects of scanning microdensitometry. I. Stray light (glare). J Microsc 92:1–16

Goldstein DJ (1971) Aspects of scanning microdensitometry. II. Spot size, focus and resolution. J Microsc 93:15–42

Gomori G (1952) Microscopic histochemistry. University Press, Chicago

Gooding PE, Chayen J, Sawyer B, Slater TF (1978) Cytochrome P-450 distribution in rat liver and the effect of phenobarbitone administration. Chem Biol Interact 20:299–310

Greep RO, van Dyke HB, Chow BF (1941) Use of anterior lobe of prostate gland in the assay of metakentrin. Proc Soc Exp Biol Med 46:644–649

Gregory RA (1969) Gastrin- the natural history of a peptide hormone. Harvey Lect 64:121–155

Gregory RA, Tracy HJ (1961) The preparation and properties of gastrin. J Physiol (Lond) 156:523–543

Gregory RA, Tracy HJ (1964) The constitution and properties of two gastrins extracted from hog antral mucosa. Gut 5:103–114

Gregory RA, Tracy HJ (1974) Isolation of two minigastrins from Zollinger-Ellison tumour tissue. Gut 15:683–685

Gregory RA, Tracy HJ, Grossman MI (1966) Human gastrin: isolation, structure and synthesis. Nature 209:583

Gumaa KA, McLean P, Greenbaum AL (1971) Compartmentation in relation to metabolic control in liver. Essays Biochem 7:39–86

Hall R, Smith BR, Muktar ED (1975) Thyroid stimulators in health and disease. Clin Endocrinol (Oxf) 4:213–230

Haning R, Tait SAS, Tait JF (1970) In vitro effects of ACTH, angiotensins, serotonin and potassium on steroid output and conversion of corticosterone to aldosterone by isolated adrenal cells. Endocrinology 87:1146–1167

Hansson HPJ (1967) Histochemical demonstration of carbonic anhydrase activity. Histochemie 11:112–128
Hashizume K, Onaya T, Sato A (1975) The role of the pentose phosphate shunt in thyrotropin-induced thyroid hormone secretion: in vivo and in vitro studies with 6-aminonicotinamide in mouse thyroids. Endocrinology 97:962–968
Häusler G (1958) Zur Technik und Spezifität des histochemischen Carbo-anhydrasenachweis im Modellversuch und Gewebsschnitten von Rattennieren. Histochemie 1:29–47
Hayano M, Lindberg MC, Dorfman RI, Hancock JEH, Doering WE (1955) On the mechanism of the C-11β-hydroxylation of steroids; a study with H_2O^{18} and O_2^{18}. Arch Biochem Biophys 59:529–532
Henderson B, Bitensky L, Chayen J (1978) Altered phospholipids in human rheumatoid synoviocytes. Ann Rheum Dis 37:24–29
Henderson B, Bitensky L, Chayen J (1979) Glycolytic activity in human synovial lining cells in rheumatoid arthritis. Ann Rheum Dis 38:63–67
Hodges JR (1976) The hypothalamo-pituitary-adrenocortical system. J Pharm Pharmacol 28:379–382
Hodges J. R., Hotston RT (1970) Ascorbic acid deficiency and pituitary adrenocortical activity in the guinea-pig. Br J Pharmacol 40:740–746
Hodges JR, Vellucci SV (1975) The effect of reserpine on the hypothalamo-pituitary-adrenocortical function in the rat. Br J Pharmacol 53:555–561
Hoile R (1979) MS dissertation, University of London
Hoile RW, Loveridge N (1976) Preliminary studies on a cytochemical section bioassay for gastrin. J Endocrinol 71:87–88 P
Hoile R, Loveridge N (1977) Early results from the cytochemical assay for gastrin. Br J Surg 64:293
Holdaway IM, Rees LH, Landon J (1973) Circulating corticotrophin levels in severe hypopituitarism and in the neonate. Lancet 2:1170–1172
Holdaway IM, Kramer RM, McNeilly AS, Rees LH, Chard T (1974 a) Applications of the redox bioassay for luteinizing hormone. Clin Endocrinol (Oxf) 3:383–388
Holdaway IM, Rees LH, Ratcliffe JG, Besser GM, Kramer RM (1974 b) Validation of the redox cytochemical assay for corticotrophin. Clin Endocrinol (Oxf) 3:329–334
Holter H (1961) The induction of pinocytosis. In: Harris RJC (ed) Biological approaches to cancer chemotherapy. Academic Press, London New York, pp 77–88
Holter H (1965) Passage of particles and macromolecules through cell membranes. Symp Soc Gen Microbiol 15:89–120
Holter H, Møller KM (1976) The Carlsberg laboratory 1876/1976. Rodos, Copenhagen
Holtzer H, Abbott J (1968) Oscillations of the chondrogenic phenotype in vitro. In: Ursprung H (ed) The stability of the differentiated state. Springer, Berlin Heidelberg New York, pp 1–16
Hornby WE, Lilly MD, Crook EM (1966) The preparation and properties of ficin chemically attached to carboxymethylcellulose. Biochem J 98:420–425
Hornby WE, Lilly MD, Crook EM (1968) Some changes in the reactivity of enzymes resulting from their chemical attachment to water-soluble derivatives of cellulose. Biochem J 107:669–674
Howard A, Pelc SR (1951) Nuclear incorporation of P^{32} as demonstrated by autoradiographs. Exp Cell Res 2:178–187
Howard A, Pelc SR (1953) Synthesis of desoxyribonucleic acid in normal and irradiated cells and its relation to chromosome breakage. Heredity (Lond) 6:261–273
Hudson AM, McMartin C (1975) An investigation of the involvement of adenosine 3′:5′-cyclic monophosphate in steroidogenesis by using isolated adrenal cell column perfusion. Biochem J 148:539–544
Huxley JS (1935) Chemical regulation and the hormone concept. Biol Rev 10:427–441
Illig R, Krawaczynska H, Torresani T, Prader A (1975) Elevated plasma TSH and hypothyroidism in children with hypothalamic hypopituitarism. J Clin Endocrinol Metab 41:722–728
Jacobs HS, Nabarro JDN (1969) Tests of hypothalamic-pituitary-adrenal function in man. QJ Med 38:475–492
Jacobson W, Webb M (1952) The two types of nucleoproteins during mitosis. Exp Cell Res 3:163–183
James VHT, Landon J, Wynn V, Greenwood FC (1968) A fundamental defect of adrenocortical control in Cushing's disease. J Endocrinol 40:15–28
Jänne J, Raina A, Siimes M (1968) Mechanism of stimulation of polyamine synthesis by growth hormone in rat liver. Biochim Biophys Acta 166:419–426
Johnson LR (1975) Trophic action of gastrointestinal hormones. In: Thompson JC (ed) Gastrointestinal hormones: a symposium. University of Texas Press, Austin, pp 215–230

Johnson LR, Grossman MI (1971) Intestinal hormones as inhibitors of gastric secretion. Gastroenterology 60:120–144

Karnovsky ML (1962) Metabolic basis of phagocytic activity. Physiol Rev 42:143–168

Keenan J, Thompson JB, Chamberlain MA, Besser GM (1971) Prolonged corticotrophic action of a synthetic substituted $^{1-18}$ ACTH. Br Med J 3:742–743

Kendall-Taylor P (1972) Adenyl cyclase activity in the mouse thyroid. J Endocrinol 52:533–549

Keutmann HT, Jacobs JW, Bishop WH, Ryan RJ (1974) Human luteinizing hormone: amino-terminal sequence analysis of the β-subunit using [^{35}S] phenylisothiocyanate. Hoppe-Seyler's Z Physiol Chem 355:935–938

Kirkham KE (1962) A new bioassay technique for the measurement in vitro of thyrotrophic hormone in serum and pituitary extracts. J Endocrinol 25:259–275

Knight CA (1967) The freezing of supercooled liquids. Van Nostrand, Princeton

Koritz S, Bhargava G, Schwartz E (1977) ACTH action on adrenal steroidogenesis. Ann NY Acad Sci 297:329–335

Kowal J (1970) ACTH and the metabolism of adrenal cell cultures. Recent Prog Horm Res 26:623–687

Kramer RM, Holdaway IM, Rees LH, McNeilly AS, Chard T (1974) Technical aspects of the redox bioassay for luteinizing hormone. Clin Endocrinol (Oxf) 3:375–381

Kramer RM, Holdaway IM, Crighton DB, McNeilly AS, Rees LH, Chard T (1976) Comparison of the redox bioassay with other assays for luteinizing hormone. J Endocrinol 69:205–211

Krebs HA, Eggleston LV (1974) The regulation of the pentose phosphate cycle in rat liver. Adv Enzyme Regul 24:422–434

Krieger DT (1974) Glandular end-organ deficiency associated with secretion of biologically inactive pituitary peptides. J Clin Endocrinol Metab 38:964–975

Krieger DT, Allen W, Rizzo F, Krieger HP (1971) Characterization of the normal temporal pattern of plasma corticosteroid levels. J Clin Endocrinol 32:266–284

Kurnick NB (1949) Methyl green-pyronin. I. Basis of selective staining of nucleic acids. J Gen Physiol 33:243–264

Kurnick NB (1950) The quantitative estimation of desoxyribonucleic acid based on methyl green staining. Exp Cell Res 1:151–158

Kurnick NB, Mirsky AE (1949) Methyl green-pyronin. II. Stoichiometry of reaction with nucleic acids. J Gen Physiol 33:265–274

La Cour LF, Deeley EM, Chayen J (1956) Variations in the amount of Feulgen stain in nuclei of plants grown at different temperatures. Nature, 177:272–273

Landon J, Wynn V, James VHT (1963) The adrenocortical response to insulin-induced hypoglycaemia. J Endocrinol 27:183–205

Landon J, Greenwood FC, Stamp TCB, Wynn V (1966) The plasma sugar, free fatty acid, cortisol and growth hormone response to insulin, and the comparison of this procedure with other tests of pituitary and adrenal function. II. In patients with hypothalamic or pituitary dysfunction or anorexia nervosa. J Clin Invest 45:437–450

Lasnitski I (1965) The action of hormones on cell and organ cultures. In: Willmer EN (ed) Cells and tissues in culture, vol 1. Academic Press, London New York, pp 591–658

Leblond CP, Percival WL, Gross J (1948) Autographic localization of radioiodine in stained sections of thyroid gland by coating with photographic emulsion. Proc Soc Exp Biol Med 67:74–76

Leblond CP, Kopriwa B, Messier B (1963) Radioautography as a histochemical tool. In: Wegmann R (ed) Histochemistry and cytochemistry. Pergamon, Oxford, pp 1–31

Leuchtenberger C (1958) Quantitative determination of DNA in cells by Feulgen microspectrophotometry. In: Danielli JF (ed) General cytochemical methods vol 1. Academic Press, New York, pp 219–278

Leuchtenberger C, Vendrely R, Vendrely C (1951) A comparison of the content of desoxyribosenucleic acid (DNA) in isolated animal nuclei by cytochemical and chemical methods. Proc Natl Acad Sci USA 37:33–38

Levine JH, Nicholson WE, Peytremann A, Orth DN (1975) The mechanism of ACTH stimulation of adrenal ornithine decarboxylase activity. Endocrinology 97:136–144

Linderstrøm-Lang K, Mogensen KR (1938) Studies on enzymatic histochemistry, 32. CR Trav Lab Carlsberg Ser Chim 23:37–42

Lipscomb HS, Nelson DH (1962) A sensitive biologic assay for ACTH. Endocrinology 71:13–23

Loeber JG, Toorenenbergen AW van, Lequin RM (1977) Some characteristics of two forms of human luteinizing hormone. J Endocrinol 72:17–18 P

Loraine JA, Bell ET (1971) Hormone assays and their clinical application. Livingstone, Edinburgh London, p 14
Lovelock JE (1957) The denaturation of lipid-protein complexes as a cause of damage by freezing. Proc R Soc Lond (Biol) 147:427–433
Loveridge N (1977) M Phil dissertation, Brunel University
Loveridge N (1978) A quantitative cytochemical method for measuring carbonic anhydrase activity. Histochem J 10:361–372
Loveridge N, Robertson WR (1978) Stimulation of adrenal 5-ene, 3β-hydroxysteroid dehydrogenase by corticotrophin in vitro. J Endocrinol 78:457–458
Loveridge N, Bloom SR, Welbourn RB, Chayen J (1974) Quantitative cytochemical estimation of the effect of pentagastrin (0.005–5 pg/ml) and of plasma gastrin on the guinea-pig fundus in vitro. Clin Endocrinol (Oxf) 3:389–396
Loveridge N, Alaghband-Zadeh J, Daly JR, Chayen J (1975) The nature of the redox change measured in the cytochemical bioassay of corticotrophin. J Endocrinol 67:28 P
Loveridge N, Hoile RW, Johnson AG, Chayen J (1978) The cytochemical measurement of gastrin-like activity. J Endocrinol 77:40–41 P
Loveridge N, Zakarija M, Bitensky L, McKenzie JM (1979) The cytochemical bioassay for thyroid-stimulating antibody of Graves' disease: further experience. J Clin Endocrinol Metab 49:610–615
Lowenstein WR (1966) Permeability of membrane junctions. Ann NY Acad Sci 137:441–472
Lowry PJ, McMartin C (1974) Measurement of the dynamics of stimulation and inhibition of steroidogenesis in isolated rat adrenal cells using column perfusion. Biochem J 142:287–294
Luyet BJ (1951) Survival of cells, tissues and organisms after ultra-rapid freezing. In: Harris RJC (ed) Freezing and drying. Institute of Biology, London, pp 77–98
Luyet BJ (1960) On various phase transitions occurring in aqueous solutions at low temperatures. Ann NY Acad Sci 85:549–569
Lynch R, Bitensky L, Chayen J (1965) On the possibility of super-cooling in tissues. J R Microsc Soc 85:213–222
Macha N, Bitensky L, Chayen J (1975) The effect of thyrotrophin on oxidative activity in thyroid follicle cells. Histochemistry 41:323–334
Mahler HR (1953) Multienzyme sequences in soluble extracts. Int Rev Cytol 2:201–230
Maier R, Barthe PL, Schenkel-Hulliger L, Desaulles PA (1971) The biological activity of [1-D-serine, 17–18–dylysine]-β-corticotrophin-(1–18)-octadecapeptide-amide. Acta Endocrinol (Kbh) 68: 458–466
Mamont PS, Böhlen P, McCann PP, Bey P, Schuber F, Tardif C (1976) α-methyl ornithine, a potent competitive inhibitor of ornithine decarboxylase, blocks proliferation of rat hepatoma cells in culture. Proc Natl Acad Sci USA 73:1626–1630
Mamont PS, Duchesne M-C, Grove J, Bey P (1978) Anti-proliferative properties of DL-α-difluoromethyl ornithine in cultured cells. A consequence of the irreversible inhibition of ornithine decarboxylase. Biochem Biophys Res Commun 81:58–66
Manley SW, Bourke JR, Hawker RW (1974) The thyrotrophin receptor in guinea-pig thyroid homogenate. J Endocrinol 61:437–448
Maren TH (1967) Carbonic anhydrase: chemistry, physiology and inhibition. Physiol Rev 47:595–781
Matsuyama H, Ruhman-Wennhold A, Johnson LR, Nelson DH (1972) Disappearance rates of exogenous and endogenous ACTH from rat plasma measured by bioassay and radioimmunoassay. Metabolism 21:30–35
McKenzie JM (1958 a) Delayed thyroid response to serum from thyrotoxic patients. Endocrinology 62:865–868
McKenzie JM (1958 b) The bioassay of thyrotrophin in serum. Endocrinology 63:372–382
McKenzie JM (1968) Humoral factors in the pathogenesis of Graves' disease. Physiol Rev 48:252–310
McKenzie JM, Zakarija M (1977) LATS in Graves' disease. Recent Prog Horm Res 33:29–57
McKerns KW (1966) Hormone regulation of the generic potential through the pentose phosphate pathway. Biochim Biophys Acta 121:207–209
McLachlan SM, Smith BR, Petersen VB, Davies TF, Hall R (1977) Thyroid-stimulating autoantibody production in vitro. Nature 270:447–449
McLaren AD, Packer L (1970) Some aspects of enzyme reactions in heterogeneous systems. Adv Enzymol 33:245–308
McPartlin J, Skrabanek P, Powell D (1977) Bone alkaline phosphatase: Quantitative cytochemical characterization and the response to parathyrin in vitro. Biochem Soc Trans 5:1734–1736

Meakin JW, Bethune JE, Despointes RH, Nelson DH (1959) The rate of disappearance of ACTH activity from the blood of humans. J Clin Endocrinol 19:1491–1495
Mehdi SQ, Nussey SS (1975) A radio-ligand receptor assay for the long-acting thyroid stimulator. Biochem J 145:105–111
Mendelsohn FA, Mackie C (1975) Relation of intracellular K^+ and steroidogenesis in isolated adrenal zona glomerulosa and fasciculata cells. Clin Sci Mol Med 49:13–26
Michaelis L (1947) The nature of the interaction of nucleic acid and nuclei with basic dyestuffs. Cold Spring Harbour Symp Quant Biol 12:131–142
Moline SW, Glenner GG (1964) Ultrarapid tissue freezing in liquid nitrogen. J Histochem Cytochem 12:777–783
Moscona A, Trowell OA, Willmer EN (1965) Methods. In: Willmer EN (ed) Cells and tissues in culture, vol 1. Academic Press, London New York, pp 19–98
Moyle WR, Macdonald GJ, Garfink JE (1976) Role of histone kinases as mediators of corticotropin-induced steroidogenesis. Biochem J 160:1–9
Mühlen A von zur, Hashimoto T, Emrich E, Döhler K-D (1977) Cytochemical TSH bioassay in plasma of patients with 'euthyroid goiter' and negative TRH-test. Acta Endocrinol [Suppl 208] (Kbh) 84:110–111
Mühlen A von zur, Döhler K-D, Poernomo J (1978) Further advancement of the cytochemical bioassay for human TSH. Acta Endocrinol [Suppl 209] (Kbh) 85:105–106
Munro DS (1977) Autoimmunity and the thyroid gland. Proc R Soc Med 70:855–857
Narumi S, Kanno M (1973) Effects of gastric acid stimulants and inhibitors on the activities of HCO_3^- stimulated Mg^{2+} dependent ATPase and carbonic anhydrase in rat gastric mucosa. Biochim Biophys Acta 311:80–89
Ney RL, Ogata E, Shimizue N, Nicholson WE, Liddle GW (1964) Structure-function relationships of ACTH and MSH analogues. Proc 2nd Int Congr Endocrinol. Excerpta Med Int Congr Ser 83:1184–1191
Niles NR, Bitensky L, Chayen J, Cunningham GJ, Braimbridge MV (1964) The value of histochemistry in the analysis of myocardial dysfunction. Lancet 1:963–965
Nureddin A (1977) Ovarian ornithine decarboxylase induction: a specific and rapid in vivo bioassay of LH. Biochem Med 17:67–79
Nureddin A, Hartree AS, Johnson P (1972) Purification and properties of human pituitary LH and its subunits. In: Saxena BB, Beling CG, Gandy HM (eds) Gonadotropins. Wiley, New York London, pp 167–173
Onaya T, Kotani M, Yamada T, Ochi Y (1973) A new in vitro test to detect thyroid stimulators in sera from patients by measuring colloid droplet formation and cyclic AMP in human thyroid slices. J Clin Endocrinol Metab 36:859–870
Orth DN, Nicholson WE (1977) Different molecular forms of ACTH. Ann NY Acad Sci 297: 27–48
Ostrowski K, Kommender J, Košcianek H, Kwarecki K (1962a) Quantitative investigation of the P and N loss in the rat liver when using various media in the 'freeze-substitution' technique. Experientia 18:142–146
Ostrowski K, Kommender J, Košcianek H, Kwarecki K (1962b) Quantitative studies on the influence of the temperature applied in freeze-substitution on P, N, and dry mass losses in fixed tissue. Experientia 18:227–230
Ostrowski K, Kommender J, Košcianek H, Kwarecki K (1962c) Elution of some substances from the tissues fixed by the 'freeze-substitution' method. Acta Biochim Pol 9:125–130
Parlow AF (1958) A rapid bioassay method for LH and factors stimulating LH secretion. Fed Proc 17:402
Parlow AF (1961) Bioassay of pituitary LH by depletion of ovarian ascorbic acid. In: Albert A (ed) Human pituitary gonadotrophins. Thomas, Springfield, pp 300–310
Parsons JA (1976) Parathyroid physiology and the skeleton. In: Bourne GH (ed) Biochemistry and physiology of bone, vol 4, 2nd edn. Academic Press, New York London, pp 159–218
Parsons JA, Potts JTJr (1972) Physiology and chemistry of parathyroid hormone. In: MacIntyre I (ed) Calcium metabolism and bone disease, vol 1. Saunders, London Philadelphia, pp 33–78
Parsons JA, Reit B (1974) Chronic response of dogs to parathyroid hormone infusion. Nature 250:254–257
Parsons JA, Reit B, Robinson CJ (1973) A bioassay for parathyroid hormone using chicks. Endocrinology 92:454–462

Parsons JA, Rafferty B, Gray D, Reit B, Zanelli JM, Keutmann HT, Tregear GW, Callahan EN, Potts JTJr (1975) Pharmacology of parathyroid hormone and some of its fragments and analogues. In: Talmadge RV, Owen M, Parsons JA (eds) Calcium regulating hormones. Excerpta Medica, Amsterdam, pp 33–39

Patel YC, Burger HG (1973) Serum thyrotrophin in pituitary and/or hypothalamic hypopituitarism. J Endocrinol Metab 37:190–196

Pattison JR, Bitensky L, Chayen J (eds) (1979) Quantitative cytochemistry and its applications. Academic Press, London

Peckham WD, Yamaji T, Dierschke DJ, Knobil E (1973) Gonadal function and the biological and physicochemical properties of follicle-stimulating hormone. Endocrinology 92:1660–1666

Pelc SR (1958) Autoradiography as a cytochemical method with special reference to C^{14} and S^{35}. In: Danielli JF (ed) General cytochemical methods. Academic Press, New York, pp 279–316

Pelc SR, Howard A (1952) Techniques of autoradiography and the application of the stripping film method to the problems of nuclear metabolism. Br Med Bull 8:132–135

Peters TJ, Heath JR, Wansbrough-Jones MH, Doe WF (1975) Enzyme activities and properties of lysosomes and brush borders in jejunal biopsies from control subjects and patients with coeliac disease. Clin Sci Mol Med 48:259–267

Petersen V, Smith BR, Hall R (1975) A study of thyroid stimulating activity in human serum with the highly sensitive cytochemical bioassay. J Clin Endocrinol Metab 41:199–202

Petersen VB, McGregor AM, Belchetz PE, Elkeles RS, Hall R (1978) The secretion of thyrotrophin with impaired biological activity in patients with hypothalamic-pituitary disease. Clin Endocrinol (Oxf) 8:397–402

Pierce JG (1971) The subunits of pituitary thyrotrophin – their relationship to other glycoprotein hormones. Endocrinology 89:1331–1342

Quazi MH, Romani P, Diczfalusy E (1974) Discrepancies in plasma LH activities as measured by radioimmunoassay and an in vitro bioassay. Acta Endocrinol (Kbh) 77:672–685

Raisz LG (1963) Stimulation of bone resorption by parathyroid hormone in tissue culture. Nature 197:1015

Ramachandran J, Rao AJ, Liles S (1977) Studies of the trophic effect of ACTH. Ann NY Acad Sci 297:336–348

Reader SCJ, Alaghband-Zadeh J, Carter GD, Daly JR (1976) Observations on the feedback regulation of adrenocorticotrophin secretion. J Endocrinol 71:57–58 P

Rees LH (1975) The biosynthesis of hormones by non-endocrine tumours-a review. J Endocrinol 67:143–175

Rees LH (1977) ACTH, Lipotrophin and MSH in health and disease. Clin Endocrinol Metab 6:137–153

Rees LH, Ratcliffe JG (1974) Ectopic hormone production by non-endocrine tumours. Clin Endocrinol (Oxf) 3:263–299

Rees LH, Holdaway IM, Kramer RM, McNeilly AS, Chard T (1973 a) A new bioassay for luteinizing hormone. Nature 244:232–234

Rees LH, Ratcliffe JG, Besser GM, Kramer RM, Landon J, Chayen J (1973 b) A comparison of the redox assay for ACTH with previous assays. Nature [New Biol] 241:84–86

Reichert LE, Ward DN, Niswender GD, Midgley AR (1970) On the isolation and characterization of subunits of human pituitary luteinizing hormone. In: Butt WR, Crooke AC, Ryle M (eds) Gonadotrophins and ovarian development. Livingstone, Edinburgh London, pp 149–154

Reynolds JJ, Dingle JT (1970) A sensitive in vitro method for studying the induction and inhibition of bone resorption. Calcif Tissue Res 4:339–349

Richman R, Park S, Yu S, Burke G (1975) Regulation of thyroid ornithine decarboxylase (ODC) by thyrotropin. I. The rat. Endocrinology 96:1403–1412

Romani P, Robertson DM, Diczfalusy E (1976) Biologically active luteinizing hormone (LH) in plasma. 1. Validation of the in vitro bioassay when applied to plasma of women. Acta Endocrinol (Kbh) 83:454–465

Romani P, Robertson DM, Diczfalusy E (1977) Biologically active luteinizing hormone (LH) in plasma. 2. Comparison with immunologically active LH levels throughout the human menstrual cycle. Acta Endocrinol (Kbh) 84:697–712

Roodyn DB (1968) The mitochondrion. In: Bittar EE, Bittar N (eds) The biological basis of medicine, vol 1. Academic Press, New York, pp 123–177

Roos P, Jacobson G, Wide L (1975 a) Isolation of five active thyrotropin components from human pituitary gland. Biochim Biophys Acta 379:247–261

Roos P, Nyberg L, Wide L, Genzell C (1975b) Human pituitary luteinizing hormone. Isolation and characterization of four glycoproteins with luteinizing activity. Biochim Biophys Acta 405: 363–379

Ross KFA (1967) Phase contrast and interference microscopy for cell biologists. Arnold, London

Russell DH, Snyder SH, Medina VJ (1970) Growth hormone induction of ornithine decarboxylase in rat liver. Endocrinology 86:1414–1419

Ryan KJ, Engel LL (1957) Hydroxylation of steroids at carbon 21. J Biol Chem 225:103–114

Sairam MR, Papkoff H, Choh HL (1972) Human pituitary interstitial cell stimulating hormone: primary structure of the α subunit. Biochem Biophys Res Comun 48:530–577

Salganik RI, Argutenskaya SV, Bersinbaev RI (1972) The stimulating action of gastrin pentapeptide, histamine and cyclic adenosine 3′5′-monophosphate on carbonic anhydrase in rat stomach. Experientia 28:1190–1191

Sayers MA, Sayers G, Woodbury LA (1948) The assay of adrenocorticotrophic hormone by the adrenal ascorbic acid-depletion method. Endocrinology 42:379–393

Sayers G, Swallow RL, Giordano ND (1971) An improved technique for the preparation of isolated rat adrenal cells: a sensitive, accurate and specific method for the assay of ACTH. Endocrinology 88:1063–1068

Scanlon MF, Smith BR, Hall R (1978a) Thyroid-stimulating hormone: neuroregulation and clinical applications. Clin Sci Mol Med 55:1–10

Scanlon MF, Smith BR, Hall R (1978b) Thyroid-stimulating hormone: neuroregulation and clinical applications. Clin Sci Mol Med 55:129–138

Schäfer HJ, Böcker W (1976) TSH-Stimulierung und zytochemische Aktivität lysosomaler Enzyme der Schilddrüse. Eine mikrodensitometrische Studie. Verh Dtsch Ges Pathol 60:206–211

Schally AV, Arimura A, Bowers CY, Kastin AJ, Sawano S, Redding TW (1968) Hypothalamic neurohormones regulating anterior pituitary function. Recent Prog Horm Res 24:497–588

Schally AV, Redding TW, Bowers CY, Barrett JF (1969) Isolation and properties of porcine thyrotropin-releasing hormone. J Biol Chem 244:4077–4088

Schams D, Karg H (1969) Radioimmunologische LH-Bestimmung im Blutserum vom Rind unter besonderer Berücksichtigung des Brunstzyklus. Acta Endocrinol (Kbh) 61:96–103

Schulster D (1974) Adrenocorticotrophic hormone and the control of adrenal corticosteroidogenesis. Adv Steroid Biochem Pharmacol 4:233–295

Schwyzer R (1977) ACTH: A short introductory review. Ann NY Acad Sci 297:3–26

Scott JE (1974) The Feulgen reaction in polyvinyl alcohol or polyethylene glycol solution. 'Fixation' by excluded volume. J Histochem Cytochem 22:833–835

Segre GV, Habener JF, Powell D, Tregear GW, Potts JTJr (1972) Parathyroid hormone in human plasma. Immunochemical characterization and biological implications. J Clin Invest 51:3163–3172

Segre GV, Niall HD, Habener JF, Potts JTJr (1974) Metabolism of parathyroid hormone. Physiologic and clinical significance. Am J Med 56:774–784

Sherman P (1955) Factors influencing emulsion viscosity and stability. Research 8:396–401

Shishiba Y Solomon DH, Beall GN (1967) Comparison of early effects of thyrotropin and long-acting thyroid stimulator on thyroid secretion. Endocrinology 80:957–961

Shome B, Parlow AF (1974) Human follicle stimulating hormone (hFSH). First proposal for the amino acid sequence of the α-subunit (hFSHα) and first demonstration of its identity with the α-subunit of human luteinizing hormone (hLHα). J Clin Endocrinol Metab 39:199–202

Siekevitz P (1962) The relation of cell structure to metabolic activity. In: Allen JM (ed) The molecular control of cellular activity. McGraw Hill, New York, pp 143–166

Silcox AA, Poulter LW, Bitensky L, Chayen J (1965) An examination of some factors affecting histological preservation in frozen sections of unfixed tissue. J R Microsc Soc 84:559–564

Simpson ER, Cooper DY, Estabrook RW (1969) Metabolic events associated with steroid hydroxylation by the adrenal cortex. Recent Prog Horm Res 25:523–562

Smith BR, Hall R (1974) Binding of thyroid stimulators to thyroid membranes FEBS Lett 42:301–304

Smith MT, Loveridge N, Wills ED, Chayen J (1979) The distribution of glutathione in the rat liver lobule. Biochem J 182:103–108

Smyth DG (1978) The common prohormone of corticotropin and the opiate peptide lipotropin C-fragment (beta-endorphin). Biochem Soc Trans 6:61–63

Snell CR (1978) The receptor conformation of flexible peptides. Biochem Soc Trans 6:138–141

Snook RB, Saatman RR, Hansel W (1971) Serum progesterone and luteinizing hormone levels during the bovine estrous cycle. Endocrinology 88:678–686

Spencer-Peet J, Daly JR, Smith V (1964) A simple method for improving the specificity of the fluorimetric determination of adrenal corticosteroids in human plasma. J Endocrinol 31:235–238
Stadil F, Rehfeld JF (1973) Determination of gastrin in serum: an evaluation of the reliability of a radioimmunoassay. Scand J Gastroenterol 8:101–112
Stagg BH, Lewin MR, Boulos PB, Clark CG (1971) The release of gastrin in response to insulin, food, and meat-extract (Oxo). Br J Surg 58:863
Staudinger HJ, Krisch K, Leonhäuser S (1961) Role of ascorbic acid in microsomal electron transport and the possible relationship to hydroxylation reactions. Ann NY Acad Sci 92:197–207
Stuart J, Simpson JS (1970) Dehydrogenase enzyme cytochemistry of unfixed leucocytes. J Clin Pathol 23:517–521
Sturman JA (1976) Effect of methylglyoxal bis (guanylhydrazone) (MGBG) in vivo on the decarboxylation of S-adenosylmethionine and synthesis of spermidine in the rat and guinea pig. Life Sci 18:879–886
Swift HH (1953) Quantitative aspects of nuclear nucleoproteins. Int Rev Cytol 2:1–76
Sylvén B (1954) Metachromatic dye-substrate interactions. Q J Microsc Sci 95:327–358
Symington T (1969) Functional pathology of the human adrenal gland. Livingstone, Edinburgh London
Tabor CW, Tabor H (1976) 1,4-diaminobutane (putrescine), spermidine and spermine. Ann Rev Biochem 45:285–306
Tappel AL (1969) Lysosomal enzymes and other components. In: Dingle JT, Fell HB (eds) Lysosomes in biology and pathology vol 2. North Holland, Amsterdam, pp 207–244
Tashjian A, Ontjes DJ, Munson PL (1964) Alkylation and oxidation of methionine in bovine parathyroid hormone; effects on hormonal activity and antigenicity. Biochemistry 3:1175–1182
Tracy HJ, Gregory RA (1964) Physiological properties of a series of synthetic peptides structurally related to Gastrin I. Nature 204:935–938
Trowell OA (1959) The culture of mature organs in a synthetic medium. Exp Cell Res 16:118–147
Tsuruhara T, van Hall EV, Dufau ML, Catt KJ (1972) Ovarian binding of intact and desialated hCG in vivo and in vitro. Endocrinology 91:463–469
Tunbridge WMG, Evered DC, Hall R, Appleton D, Brewis M, Clark F, Grimley Evans J, Young E, Bird T, Smith PA (1977) The spectrum of thyroid disease in a community: the Whickham survey. Clin Endocrinol (Oxf) 7:481–493
Van Damme MP, Robertson DM, Romani P, Diczfalusy E (1973) A sensitive in vitro bioassay method for luteinizing hormone (LH) activity. Acta Endocrinol [Suppl] (Kbh) 177:166
Van Damme MP, Robertson DM, Diczfalusy E (1974) An improved in vitro bioassay method for measuring luteinizing hormone (LH) activity using mouse Leydig cell preparations. Acta Endocrinol (Kbh) 77:655–671
Van Hell H, Matthijsen R, Overbeek GA (1964) Effects of human menopausal gonadotrophin preparations in different bioassay methods. Acta Endocrinol (Kbh) 47:409–418
Vendrely R, Vendrely C (1948) La teneur du noyau cellulaire en acide désoxyribonucléique à travers les organes, les individus et les espèces animales. Experientia 4:434–436
Volpé R, Farid NR, Westarp C von, Row VV (1974) The pathogenesis of Graves' disease and Hashimoto's thyroiditis. Clin Endocrinol (Oxf) 3:239–261
Walker PMB (1958) Ultraviolet microspectrophotometry. In: Danielli JF (ed) General cytochemical methods. Academic Press, New York, pp 163–217
Walker PMB, Yates HB (1952) Ultraviolet absorption of living cell nuclei during growth and division. Symp Soc Exp Biol 6:265–276
Walsh JH, Grossman MI (1975) Gastrin. New Engl J Med 292:1324–1334
Walsh JH, Maxwell V, Isenberg JI (1975) Biological activity and clearance of human big gastrin in man. Clin Res 23:259 A
Watson J (1972) Plasma levels of luteinizing hormone in women and rats measured by bioassay. J Endocrinol 54:119–123
Webster LA, Atkins D, Peacock M (1974) A bioassay for parathyroid hormone using whole mouse calvaria in tissue culture. J Endocrinol 62:631–637
WHO Expert Committee on Biological Standardization (1975) 26th report. WHO Tech Rep Ser 565
Willmer EN (1965 a) Morphological problems of cell type, shape and identification. In: Willmer EN (ed) Cells and tissues in culture, vol 1. Academic Press, London New York, pp 143–176
Willmer EN (ed) (1965 b) Cells and tissues in culture vols 1 and 2 Academic Press, London New York
Wilson LD, Oldham SB, Harding BW (1968) Cytochrome P-450 and steroid 11β-hydroxylation in mitochondria from human adrenal cortex. J Clin Endocrinol 28:1143–1152

Winand RJ, Wadeleux PA, Etienne-Decerf J, Kohn LD (1976) Biochemical basis of thyroid stimulation using thyroid cells in tissue culture. In: Von zur Mühlen A, Schleusener H (eds) Biochemical basis of thyroid stimulation and thyroid hormone action. Thieme, Stuttgart, pp 1–23

Wollman SH (1969) Secretion of thyroid hormones. In: Dingle JT, Fell HB (eds) Lysosomes in biology and pathology vol 2. North Holland, Amsterdam, pp 483–512

Wolman M (1955) Problems of fixation in cytology, histology and histochemistry. Int Rev Cytol 4:79–102

Yalow RS (1974) Heterogeneity of peptide hormones. Recent Prog Horm Res 30:597–633

Yalow RS, Berson SA (1970) Size and charge distinctions between endogenous human plasma gastrin in peripheral blood and heptadecapeptide gastrins. Gastroenterology 58:609–615

Yalow RS, Berson SA (1971) Further studies on the nature of immunoreactive gastrin in human plasma. Gastroenterology 60:203–214

Yalow RS, Berson SA (1972) And now 'big big' gastrin. Biochem Biophys Res Commun 48:391–395

Zanelli JM, Parsons JA (to be published) Bioassay of parathyroid hormone – a review. In: Kuhlencordt F (ed) Klinische Osteologie. Springer, Berlin Heidelberg New York. Handbuch der inneren Medizin, vol VI/I

Zanelli JM, Lea DJ, Nisbet JA (1969) A bioassay method in vitro for parathyroid hormone. J Endocrinol 43:33–46

Subject Index

Other Volumes in This Series:

Volume 16: J. E. A. McIntosh, R. P. McIntosh
Mathematical Modelling and Computers in Endocrinology
1980. 73 figures, 57 tables. XII, 337 pages
ISBN 3-540-09693-0

Volume 15: A. T. Cowie, I. A. Forsyth, I. C. Hart
Hormonal Control of Lactation
1980. 64 figures, 7 tables. Approx. 320 pages
ISBN 3-540-09680-9

Volume 14: J. H. Clark, E. J. Peck, Jr.
Female Sex Steroids
Receptors and Function
1979. 116 figures, 18 tables. XII, 245 pages
ISBN 3-540-09375-3

Volume 13: H. F. DeLuca
Vitamin D – Metabolism and Function
1979. 14 figures. VIII, 80 pages
ISBN 3-540-09182-3

Volume 12
Glucocorticoid Hormone Action
Editors: J. D. Baxter, G. G. Rousseau
1979. 176 figures, 58 tables. XIX, 638 pages
ISBN 3-540-08973-X

Volume 11: S. Ohno
Major Sex-Determining Genes
1979. 34 figures, 6 tables. XIII, 140 pages
ISBN 3-540-08965-9

Volume 10: W. I. P. Mainwaring
The Mechanism of Action of Androgens
1977. 12 figures, 17 tables. XI, 178 pages
ISBN 3-540-07941-6

Springer-Verlag
Berlin
Heidelberg
New York

Volume 9: R. E. Mancini
Immunologic Aspects of Testicular Function
1976. 36 figures, 8 tables. IX, 114 pages
ISBN 3-540-07496-1

Volume 8: E. Gurpide
Tracer Methods in Hormone Research
1975. 35 figures. XI, 188 pages
ISBN 3-540-07039-7

Volume 7: E. W. Horton
Prostaglandins
1972. 97 figures. XI, 197 pages
ISBN 3-540-05571-1

Volume 6: K. Federlin
Immunopathology of Insulin
Clinical and Experimental Studies
1971. 53 figures. XIII, 185 pages
ISBN 3-540-05408-1

Volume 5: J. Müller
Regulation of Aldosterone Biosynthesis
1971. 19 figures. VII, 137 pages
ISBN 3-540-05213-5

Volume 4: U. Westphal
Steroid-Protein Interactions
1971. 144 figures. XIII, 567 pages
ISBN 3-540-05312-3

Volume 3: F. G. Sulman
Hypothalamic Control of Lactation
In collaboration with M. Ben-David, A. Danon, S. Dikstein, Y. Givant, K. Khazen, J. Mishkinsky-Shani, I. Nir, C. P. Weller
1970. 58 figures. XII, 235 pages
ISBN 3-540-04973-8

Volume 2: K. B. Eik-Nes, E. C. Horning
Gas Phase Chromatography of Steroids
1968. 85 figures. XV, 382 pages
ISBN 3-540-04277-6

Volume 1: S. Ohno
Sex Chromosomes and Sex-Linked Genes
1967. 33 figures. X, 192 pages
ISBN 3-540-03934-1

Springer-Verlag
Berlin
Heidelberg
New York